A Practical Guide to the Safety Profession

A Practical Guide to the Safety Profession

The Relentless Pursuit

Jason A. Maldonado CSP, SMS

CRC Press
Taylor & Francis Group
Boca Raton London New York

CRC Press is an imprint of the
Taylor & Francis Group, an **informa** business

CRC Press
Taylor & Francis Group
6000 Broken Sound Parkway NW, Suite 300
Boca Raton, FL 33487-2742

© 2020 by Taylor & Francis Group, LLC

CRC Press is an imprint of Taylor & Francis Group, an Informa business

No claim to original U.S. Government works

Printed on acid-free paper

International Standard Book Number-13: 978-0-367-34749-9 (Hardback)

Library of Congress Cataloging-in-Publication Data

CIP data goes here

Visit the Taylor & Francis Web site at
www.taylorandfrancis.com

and the CRC Press Web site at
www.crcpress.com

For Nick,
I'll always miss our talks and strategy sessions.
I learned more from you about airplanes than
I probably ever wanted to know. You left a void that
can never be filled, my friend. Thanks for the stories.

CONTENTS

PART III DO WHAT MAKES A DIFFERENCE

ACKNOWLEDGMENTS

I always find it entertaining to watch celebrities give their tear-filled thanks to all of the people who have made it possible to stand on this or that prestigious stage. Mostly because those speeches feel genuine, sometimes because they're awkward. If it makes it easier, feel free to envision me in my sparkliest tiara as I look upon a crowd of adoring fans and deliver this obligatory thanks to those who have supported me in this journey. In reality, though, it's not nearly that glamorous. Those who have supported me through this process are the ones who deserve an award, not me.

I've said since I was a child that I wanted to be a writer. But the cold hard truth is that I never had the guts to pick up my pen (metaphorically speaking) and write something for the world to read. Well, now I have. But not because I'm particularly wise or worthy to teach the lessons within this book. Ultimately, these words are here because my dad finally told me to quit talking and start writing. For that I owe him an eternal debt of gratitude.

Once I started, however, there was one person who kept my feet to the fire in spite of my constant, malicious self-sabotage. That person is my wife Jennifer. She forced this thing out of me. For that, I will never be able to thank her adequately. She has endured the ups, downs, and everything in between just to see me realize a personal dream. I'm certain there's not another soul on earth who would have put up with all that (seriously, I'm a whiny little bitch sometimes).

In addition to my wife and my dad, I would be remiss if I didn't thank Phil La Duke, Clive Lloyd, Mike Rogers, Rich Marr, and Don King. None of you had any reason to give me the time of day, but your feedback has been invaluable. Thank you.

I hope you all enjoy these stories.

FOREWORD

Anyone who has sat through the excruciatingly boring time waster that passes for a safety meeting will love this book. Like Jason, I have been told too many times that safety isn't funny, and I am finally ready to admit that they are right: safety isn't funny. It is a humorless world where one wrong word can get you in trouble and chastised like the wiseacre sent to the proverbial principal's office for cracking jokes while other people are trying to learn. But it shouldn't be drudgery to reduce risk; in fact, Jason proves safety can be funny without endangering lives.

Jason takes the reader down the rabbit's hole of worker safety where good, reasonable, hardworking safety professionals get caught up in a nightmarish world performing tasks that "just don't matter." From the madness of measuring nothing, to the quirks and foibles of people who have risen far above their competency level, to people who rabidly argue for or against theories that are completely pointless. Jason brings this world to life with anecdotes from his storied career. Even if you don't work in the field of safety you will likely be able to relate to Jason's stories about the people he has met and will likely gain insight from his sage advice that, if you look hard enough, you will be able to apply in your lives.

Oh, and he drew a snake.

Phil La Duke
Global Safety Consultant
Blogger
Author of *I Know My Shoes Are Untied, Mind Your Own Business! An Iconoclast's View of Safety* and *Lone Gunman: Rewriting the Handbook on Workplace Violence Prevention*
Social Maladroit

PROLOGUE: I DREW A SNAKE ...

"Jase, safety's not funny!"

Nick's thinly veiled grin betrayed his otherwise stern chastisement. Tony blushed a little and chuckled softly with his head down. Kelly, unsurprisingly, was oblivious and completely missed the blatant sarcasm. I smiled as I put down my masterful artwork and quickly read through my non-update.

The four of us were sitting around the "Safety Table" for our daily team meeting. We had met at that table every day at 11 a.m. for the past two years. It was a routine, structured, efficient way for the Safety Department to communicate. It was also mind-numbing and excruciatingly painful. The kind of meeting that makes you wish you could dig your eyeballs out of your skull with a spoon and stuff them in your ears so you neither had to see nor hear what was going on. It was robotic:

- **11:00** Nick (our manager) would start us off by running through pertinent announcements (quick and painless).
- **11:05** Tony (second most senior to Nick) would discuss a few items from his contracts that he needed advice or input on (also quick, yet efficient).
- **11:10** Nick, Tony, and I would drift into a deep, thoughtless trance while Kelly talked through every nuance of his day. Coffee with the contractors ... how bad the Porta-Johns were ... every chemical that was going to be used on site that day ... the arguments he'd had with field engineers ... how much concrete was being poured on level 2, section five

of the new garage ... The missing period on one sentence of a contract submittal that made him question his decision to approve it ... EVERYTHING!

- **11:55** Nick, upon waking from his eyes-open nap, would ask me what I had to contribute.

That day I had only a hand-drawn picture of a snake. It was an excellent snake. Magnificent even. Thus, I shared it and was instructed ... once again ... that safety was *not* funny.

Don't get me wrong, the four of us were a high functioning, well-balanced team. Probably the best I've ever been a part of. Nick was a coach, mentor, and moral compass. Tony was an infinite databank of Occupational Safety and Health Administration (OSHA) regulation and interpretation. Kelly, in spite of lacking the ability to sense sarcasm, was tolerable compared to many more I would meet in years to come. He was unwaveringly old-school, set in his ways, and an unrelenting people pleaser (more on that later), but he did understand the nuts and bolts better than most. I drew up the rear, soaking in the knowledge of my more experienced peers while organizing and broadcasting all of the data, reports, and trends they developed in the field. After two years of daily meetings, we could almost read each other's thoughts. Truthfully, the meetings were not needed at that point, but we kept them because they had been the backbone of our success.

Nick had known from the start that the success of our fledgling safety department was entirely dependent on two things: Cohesive teamwork and clear direction. He had recognized the strengths and weaknesses in each of us and saw an urgent need to use them to create harmony rather than allow them to tear us apart and serve our own interests and career aspirations. In doing so, he bolstered each of us far beyond what we could have become as the independent operators we had been at the start. Along the way, we learned some very powerful lessons and saw some rare glimpses of what true success in industrial safety looks like. My goal is that I'll be able to do those lessons justice in the pages to come.

The day I drew the snake is seared into my mind for eternity. Not because the day in itself was significant. I honestly don't remember anything else about that day except saying "I drew a snake." The reason I remember it is because that phrase became a running gag between Nick and me until the day he passed away. Tony always chuckled when anyone mentioned it. Kelly never understood it was a joke, so I took every opportunity to snarkily throw it into conversations with him. It was a dividing line, a stark representation of the small difference between those who "get it" and the others who are wandering aimlessly in the dark clinging to catchphrases and flavors of the month. The truth is safety *can be* funny. It's deadly serious, but sometimes it's OK to laugh, especially at ourselves. Nick knew it, Tony still knows it, and I'm here at this keyboard trying to string together thoughts that (hopefully) will guide a new generation of leaders to think differently than the Kellys of the world and make real impacts in the lives of those who work with and around them.

This book is a tribute. Nick, as you may have gathered, was a wise old man (69 years old when I first met him). He took a young, ambitious bulldog who thought rules and regulations were king and made him (me) question everything he thought. I hope I can do the same for anyone reading this. Maybe it will help someone else pursue #relentlesssafety...

ABOUT THE AUTHOR

 **Jason Maldonado** has spent more than 15 years working in safety and health roles throughout a variety of industries. Beginning as an explosive safety technician while serving in the United States Air Force, Jason developed a passion for teaching the importance of working safely and helping others figure out how to accomplish that mission. After transitioning from the military, he spent time in heavy civil construction, chemical weapon demilitarization, electrical transmission and distribution, and food manufacturing. All of these experiences have helped shape his unique perspective of the state of the modern safety professional and ignited a passion to make a positive change. In addition to this book, you can read his weekly candid observations about safety at https://relentlesssafety.com. He holds the credentials of Certified Safety Professional (CSP), Safety Management Specialist (SMS), Certified Occupational Hearing Conservationist (COHC), and Certified Reliability Leader (CRL). Jason lives in Roswell, New Mexico with his wife Jennifer and their two children Anthony and Emily.

PART I

FORGET THE TRIVIAL
An Unconventional Introduction to Occupational Safety

You'll either love me or hate me after reading through this part. Almost nothing in my experience (except politics) has ever elicited as much hate, venom, and vitriol as the subjects I'm going to glaze over in the following pages. And "glaze over" is putting it mildly. I passed over some comedic gold when making the final cuts regarding what I would include here. If comedy was the sole purpose, well let's just say the "snake incident" was one of my milder antics.

In preparation for this project, I wrote out a list of stories from the span of my career, safety-related or not. My first goal was to pick out the ones that I felt would best illustrate why so many seem to just instinctively hate the "Safety Guy." I don't have any illusions of wholesale changing that perception by writing this book, but I do believe much of that negative image is self-inflicted. Those stories are what you will find in this part.

There is a huge rift in the safety community these days, filled with debates about what does and does not work. If you peruse LinkedIn or any other site that features safety forums, you'll find hordes of keyboard warriors screaming in CAPS LOCK what they know better than everyone else (whatever that may be). While I find some of it entertaining, I rarely take much away that is useful to the people I support. I truly believe that if we are to become better, dare I say even worthy of the title "Safety Professional," we have to quit spinning our wheels and do some hard self-reflection. The first step? Accepting that much of what we do makes no discernible difference. Fair warning, this isn't for the easily offended.

1

F ... ORGET OSHA

Legal disclaimer: The title of this chapter should be read as a jest, not disparagement. In no way does the author advocate that any person ignore Occupational Safety and Health Administration (OSHA) guidance or dismiss it in any way. The contents of this chapter are solely the opinion of the author and intended only for entertainment and educational purposes. These views do not represent any legal guidance or the views of any professional organization.

Now that we've addressed that, let me tell you what you're in for. If this book is to serve any purpose other than just my own catharsis, there needs to be a clear direction. So, here's what you're going to get: I'm going to tell you stories. These "stories" are nothing more than moments in time. For me, they have great significance and have helped structure my belief system. I'm hoping they will do the same for you. Or at the very least prompt you to analyze and ask questions that will help you improve yourself, your organization, and the safety of those you support. But don't stop reading if you don't work in the safety field. That's not a prerequisite. The lessons that follow are good for any aspect of management. If you don't believe me, just indulge me a little. Try some of the ideas you find in these pages. If you don't like them or they don't work (they do though), you can always go back to doing things the boring old way.

Let's play with the idea of clear direction. First, we need to discuss your target. Or, more specifically, your organization's target. If we don't get this out of the way at the start, nothing else will make sense.

Think about it. What is the first thing that comes to mind when you are asked to describe how well a company is doing with their safety program? It's not an absolute, but most organizations define their safety success by a number, a "rate." Maybe you use your Recordable Injury Rate or DART Rate, or maybe you're focused on your Lost Time Injury Rate. They're all the same measure though, so let's just say "rate" to keep it simple. The good ol' "(# of incidents x 200,000)/hours worked."

What does that tell us? While it's true that the measure demonstrates a representation of how many employees per 100 are (or are not) injured, it does not tell us *how* anything has come to be. The rate's ability to diagnose the condition of a safety culture is utterly non-existent. When used as a gauge to determine cause or severity it is entirely arbitrary. So (nod along with me here), every time we use the rate as a representation of performance, we only do so by making the weak argument that: good = a low number or no number at all. Meaning no one was injured. That's right folks, our go-to for describing safety performance is a measure that defines excellence by stating that nothing happened. If you ask me that's an insult to all of the hard work (and the people who do it) that makes your business run. I get that it's a bad deal when 5, 10, 15, people a year are injured and need medical attention. I would not wish that on anyone. But how does patting yourself on the back for a low rate of 0.36 indicate anything other than just pure luck?

Not convinced? What if I state it another way? Let's say your company ends its year with a record-breaking rate of 2.0, lowest ever for your organization, and top quartile for your industry. What will your "goal" for the following year be? A 25% year over year reduction? A "zero accident" culture? And what happens when you reach *that* mark? Have you now become a "world class" organization? Many would emphatically answer yes, and it's tempting, to be sure. But then how will you explain the "negative trend" when the organization experiences a catastrophic event that sends five people to the hospital? The problem is, when you measure nothing, you've not truly proven that you operate to any standard at all, good or bad.

Keeping with that line of thinking, there's a really practical reason to change the way we measure safety performance. How many times have you sat in a staff meeting or other venue where "safety" was the first talking point on the agenda? You may have even been the one making the report which is short, concise, and completely lacking in value: "No new incidents to report." The meeting can then progress with updates from the departments who actually have something to report. Safety reports are given that way because safety professionals have incorrectly taught our leadership that being safe means not getting hurt. If you ask me, that's a slight to anyone who has ever used their hands to make a living. Being safe doesn't mean you simply avoid an injury. It means that you put in work and do it in a way that *ensures* you don't get hurt. It's a shame we don't talk about that work and how it is accomplished.

There are many great authors who have explained at length why you cannot define safety performance by the absence of injury (read any book by Dr. Todd Conklin, for instance). So, I won't belabor that point. What I will say is that we lean on the rate because no one has come up with anything better. No one seems to want to try either. Think about it: Our lack of providing an alternative has bred a culture where we strive for the arbitrary rather than the excellent. It is truly a disservice to the people whose safety we're here to ensure. Let me illustrate that point by telling a story that likely every safety professional has either lived through or at least heard about.

One day at my facility an employee tripped and fell, rolling his ankle and falling onto his (previously injured) knee. I arrived shortly after to assess the employee and, along with his supervisor, determined that we should transport him to the facility medical provider to get checked. The fear was that he had broken the ankle or re-injured the bad knee. The department manager came in following that decision and asked me what was going to happen. I told him that he would most likely need X-rays. Beyond that, the medical provider would be the one making the determinations. Can you guess what his first question was?

"Will the X-rays make it an OSHA Recordable?" I took a deep breath and resisted the urge to make *him* an OSHA Recordable.

"No. No, they will not, but let's get him better before we worry about that." Privately I thought "What difference does it make?" I may have just skimmed the surface of the point in the earlier paragraphs, but to reiterate the point about rates, let's dig in a little. What makes an injury an "OSHA Recordable?" If you're new to this language, OSHA Recordables are those numbers of injuries per 100 employees that actually make up the rates. It should be safety 101, but it never ceases to amaze me how many safety professionals don't know what is or isn't recordable. I could probably write volumes, play out hundreds of scenarios, and even theorize quite a bit about this topic, but for the sake of keeping you awake, I'll give the cliff notes. Recordable injuries are those work-related injuries which involve death, loss of consciousness, work-related diseases such as cancer, days away/restricted work/ job transfer (commonly referred to as DART), broken bones/ teeth, needlestick/sharps injuries, and those work-related injuries that require "treatment beyond first aid." That last one is where things get stupid.

"What is treatment beyond first aid?" you ask. Well, if we ever happen to meet over some beers and you want to talk about some really nerdy stuff, we can get into it. Most safety practitioners know that OSHA defines first aid as any one of 14 very specific treatments which have no medical basis or even logical thought associated with them. For example, giving an employee an Ace-Bandage is considered first aid. However, giving that employee a rigid wrist brace would be considered medical treatment. It makes no sense.

That's it. As far as the law is concerned, if an employee is injured at work and treatment or restrictions are provided (by *anyone*) that isn't on the very specific list I just alluded to, it is recordable "for recordkeeping purposes." That statement is key because it tells you what OSHA is actually after. See, the recordkeeping standard is a "no-fault policy." It is a method of gathering data to identify trends, not a method used to indict or cast blame. Take the following actual OSHA quote for example:

> The concept of fault has never been a consideration in any record keeping system of the U.S. Department of Labor. Both the Note to Subpart A of the final rule and the new OSHA Form 300 expressly state that recording a case does not indicate fault, negligence, or compensability. In addition, OSHA recognizes that injury and illness rates do not necessarily indicate a lack of interest in safety and health or success or failure per se. OSHA feels it is to the benefit of all parties to go beyond the numbers and look at an employer's safety and health program.
>
> *(OSHA, 2002)*

Even OSHA recognizes that rates are not a good measure of performance. They're simply a tool by which comparisons are made and analyses are performed. The recording criterion is an arbitrary thing meant only to draw a line in the sand. A recordable injury is just a blip on the radar that needs to be identified and investigated, but it isn't a measure of performance. And yet the rate is what we cling to. Safety "professionals" spend eons debating the gray areas between what they think is or isn't or might be recordable rather than taking the time to move past that one instance and make positive changes in the lives of the people we supposedly support. Rates become what we base our decisions, directives, and incentives on. Yet we know that they are arbitrary.

Some will argue that, while the rate is not a "great" measurement tool, its inherent benefits outweigh its limitations. Well, folks, that's crap. It's just plain laziness on the part of number jockeys who don't want to put in the work to try and blaze a path into uncharted territory. The *only* benefits an injury rate has is to the organization, not to its people. It provides an "apples-to-apples" comparison about how many people the company hurts (or doesn't). That comparison helps the company obtain a better premium on workers' compensation rates and/or be more competitive for future work. But it doesn't measure the quality of your safety process in any way, shape, or form. It certainly doesn't empower people to play their part because they have zero control over it.

I'm not at all advocating that we stop looking back at the events that occur within our organizations. We need to learn those lessons. But there has to be a better way to analyze the past than some arbitrary, nonsensical rate. I'll get to that in Part III, but until then, I'll conclude this chapter with one final question. If the basis for your decisions about how well your safety program performs is arbitrary, what does that say about the decisions you make to determine how to guide your safety program?

We can and should do better than that. So, as I said, F ... orget OSHA.

2

HEY SAFETY GUYS! YOU MISSED A HAZARD!

The last chapter may seem overly critical of OSHA, but consider this: One of the earliest experiences I had in civilian safety life was an OSHA ten-hour course conducted by a current OSHA compliance officer. Before even beginning the required OSH Act curriculum or delving into war stories from his compliance repertoire, he challenged us all to do more than just compliance. In his own words, OSHA compliance was equivalent to getting a "D" in school. Passing, but barely. Certainly nothing to write home about. That charge from OSHA's own mouthpiece set me on a path to seek something better. Over the years, I've learned that while there is no silver bullet to safety success, there may well be a perfect starting place. To get there, we need to determine what is important and what is not.

Here you will learn about my first nemesis. I wouldn't want to tarnish any reputations, so I'll refer to him as "Francis" because it was funny when Deadpool did it. Francis was a "wonderful" human. The kind of guy who I would imagine loved to do evil things like punch kittens for fun then sit quietly at home every night crying because he couldn't figure out why his children didn't love him anymore. He was a pompous, ego-driven megalomaniac who had risen to the level of middle management not through skill, but through politicking and overpriced degrees. He believed he had POWER and he wanted you to know it.

Francis was one of four site managers at a chemical plant I had been assigned to. The plant was a joint venture between four prominent corporations, and each had staked a claim to their fair share by appointing the highest-ranking member of their respective organization over their slice of the pie. It was a constant, soul-sucking political power struggle and I had the privilege of getting in on the ground floor. His department (though he had absolutely no background in the field) was Environmental Compliance. He, along with the Engineering Manager, Operations Manager, and General Manager quickly became THE FOUR mammoth obstacles to my success as the plant technical writer for the operations safety team. The others were reasonable people (mostly), but Francis … Francis was a certified peach.

My first assignment at this new project was to research and write all of the plant safety documents and Standard Operating Procedures (SOPs). It was something I took a great amount of pride in and accomplished with a great deal of praise from my superiors. If anything, it was an assignment that bolstered my self-confidence and continued me on the path Nick had set me on. He and I still spoke regularly, but as I had transferred from the project where he had been my manager, he was now only my chief advisor. Nick had pulled some strings to get me on the crew of this new plant and I was one of the first to join the team. My new supervisor Dave was a rising star within our giant firm and gave me the freedom to develop at whatever pace I chose. All was well. For a time.

As the plant finished the final stages of construction, operations began to slowly shift into gear. At that point, I had spent nearly a year writing. Now complete, I had to present my work to THE FOUR. Thus far I had been sheltered from the contentious politics of the site as I had researched, interviewed, walked-down, pondered, and finally put pen to paper and constructed my brilliant literary directives. Then, as had been agreed upon long before my arrival, the approval process began. I could never have conceived the level of unabashed waste that would comprise that bureaucratic nightmare.

Every author needs an editor. Every good procedure needs to be reviewed. Those are not issues worth debating. I may have a heightened level of confidence in my ability to write well, but even in my most arrogant moments, I've never believed that my words have the ability to move mountains. I've met plenty of safety "pros" over the years who believe they know better than everyone else and try to legislate their way to a safe site, but if I learned anything from Nick it was the fact that there's always more to learn. So (sometimes in spite of myself, and not always successfully), I've always tried to distance myself emotionally from the words I put to paper. That's not to say that I'm not exceedingly confident in what I write, though.

Approval should have been a simple formality. All that was required was that the procedures and SOPs be reviewed by those who would be using them and then approved by THE FOUR. Easy. The plant had actually set up this process quite well and had even employed a "technical editor" who would facilitate the process. Her name was Stacy. She was a brilliant editor and a skilled writer herself. She taught me many things about words. Some of which I'll cover later.

The directions were as clear as they could be given that the plant was a government-run facility. Reviewers were to read the document and provide feedback if there were concerns about (1) safety, (2) regulatory compliance, (3) operability, (4) omissions, or (5) needed additions. Grammar and writing style were explicitly *not* supposed to be commented upon. Those items were Stacy's responsibility. Simple, right? Not for Francis. He was obstinate.

Regardless, I set out. The way it worked was that Stacy would collect my completed document which had already been reviewed by my supervisor and department manager. She would then send it to THE FOUR electronically on official letterhead. Her message included the clear instructions I laid out above and also a seven-business-day deadline for comments. They were to use Microsoft Word, turn on Track Changes, and add comments. What they actually did was turn on tracking and then proceed to bleed red lines, (incorrect) grammar corrections, and alternate word choices ... along with some completely asinine suggestions that only the elite who sat upon the leather thrones of the administration building would ever dream of thinking relevant.

At the end of the seven days (or really whenever they felt like responding) Stacy would gather all of the comments into a consolidated form and return them to me to answer. I could either accept their comments or deny the suggested changes provided I had good reasoning. Ordinary reviewers were satisfied with these rules and the responses I gave. If I denied the comment, word choice, or other suggestion, I did so because my objective was to maintain the document's usefulness, not turn it into a George R. R. Martin novel. They were intended to be concise, direct instructions. Not graphic descriptions and a rationale for every comma, ellipsis, or quotation mark along with accompanying regulatory citations.

I'm sure many have dealt with this type of ego-driven tennis match, but I'm equally sure those similar stories don't end like this one, so keep reading for the twist. I learned quickly that each of THE FOUR felt a need to demonstrate their extreme intelligence and make their mark on every aspect of the plant's operations. This was as much a power struggle as it was akin to a dog lifting its leg and marking its territory. They questioned EVERYTHING. Often betraying that hidden self-serving motivation by actually pointing out their ignorance rather than adding value. At first, I took the comments in stride. As I mentioned, I prided myself in "not being emotionally attached" to my words. But in truth, every successive round of comments and responses took its toll. The ultimate problem with the process was not that its design was flawed, rather that the ones executing it had significantly less authority than those who were expected to abide by it. In the end, what was supposed to be a simple back and forth (comment/ answer) became a constant volley only completed when *they* were satisfied.

Having little patience for that type of unwarranted ego, I would often make the obstacles much harder for myself than they needed to be (to tell the truth, I still do that ... call it a character flaw). I used a trick I had learned years earlier when working in the safety field as a junior non-commissioned officer in the US Air Force. There, my responsibilities often dictated that I had to give direction to those who were far superior in rank than I. I would spend hours crafting seemingly innocent, yet blatantly backhanded emails to identify the shortcomings of those superiors while maintaining plausible deniability when accused of being disrespectful. In my time at the chemical plant, I had nothing better to do than perfect that craft.

And so, after a satisfying year of true accomplishment, I dug in and waged literary war with THE FOUR if for no other reason than to entertain myself (just for the record, that's not a practice I recommend you implement). In time I developed a mutual respect with three of them. But not Francis. He had an unrelenting need to prove he was the best, especially in areas where he felt inadequate. Truth be told, I did too at that point in my career. He may have had some talent (though for what,

I couldn't begin to guess), but there was one thing he was certainly not as good at as I was: words.

The "comment wars" became the stuff of legend at that plant. Stacy and I led a silent coup against the tyranny that was THE FOUR, and we gained near cult status among our peers and even some of the lower level site management in doing so. That actually made things worse in terms of our ability to get documents approved, but it was fun. I always responded professionally, but secretly I would write the real responses I *wanted* to send. Table 2.1 is an example of those musings.

TABLE 2.1
Author response to the official Plant Vehicle Safety procedure

Item	Paragraph No.	Comment by Reviewer	Resolution by Originator
1	3.2.1	This paragraph discusses HR responsibilities regarding new hires and only indicates that they will verify they have a valid drivers license. Shouldn't this new hire check also include a motor vehicle history check to see if there is a history of moving violations or convictions? Drivers with poor motor vehicle history will contribute to a higher risk level in using site vehicles.	The responsibility was written based on input from HR. While it is true that personnel with bad driving records may be an increased risk, this procedure is not intended to set any subjective terms that may violate a person's rights.
2	3.7.1	This paragraph requires employees to report only alcohol and drug related traffic offenses. The site security plan requires reporting of all serious citations, including careless and reckless driving, etc. These	Reporting section was changed to reference the site security plan.

(Continued)

TABLE 2.1(Cont.)

Item	Paragraph No.	Comment by Reviewer	Resolution by Originator
		procedures should be consistent, plus if there is an indicator of poor vehicle operations, then the site should be proactive in managing this risk.	
3	4.1.7	This paragraph indicates that additional training will be required following multiple work related or at-fault vehicle incidents. How many incidents will be allowed before operating privileges are revoked?	I'll let you know the next time you're involved in an incident.
4	4.2.5	This paragraph should also include a provision that any vehicle shall be pared off the main travel path and in an area that does not interfere or obstruct other traffic and is safe for a vehicle to be parked in.	Umm. OK.
5	5.1.3	This paragraph indicates that operators are to check that all safety equipment is in place and conforms to this standard. The only requirement in this document is that a fire extinguisher be in the vehicle. Are there other requirements for a first aid kit, flares or warning devices, jacks, shovels, etc.? If so, there should be a list of those requirements in this section so that all the other	Required safety equipment is listed in appendix 02. Shovels will not be required until such time as management decides that personnel are responsible for digging fox holes in an effort to protect themselves from rabid donkeys or a zombie apocalypse occurs and employees need to carry weapons that will be used for dispatching said zombies while remaining as quiet as possible

(Continued)

TABLE 2.1(Cont.)

Item	Paragraph No.	Comment by Reviewer	Resolution by Originator
		related documents don't need to be searched for compliance with this procedure.	so as not to alert the other undead.
6	5.2.2	What is the required size and capability of the required fire extinguisher? I.e. 10A:5BC??	The one you i.e.'d sounds good.
7	Section 5.2	Should there also be a prohibition from parking vehicles in areas where vegetation is taller than 10 inches to avoid under vehicle fires caused by the muffler and catalytic converter contact?	Not unless we require employees to carry a ruler with them at all times.
8	6.1.1	This paragraph states that POVs must comply with all installation requirements. Does that include the fire extinguisher and safety equipment, backup alarms, and routine inspections? If so, then verifying POV compliance may become an issue. If not, then exclusions should be included in this paragraph.	Yes. Go get your car fixed so that it is compliant. It would take too much energy to list every possible scenario, exclusion, caveat, or clarification to keep you from misinterpreting this procedure.
9	6.3.3	This paragraph requires full time personnel to have a windshield decal on POVs. Currently there is the option for placard or decal. If this is a change in policy, then general notifications will need to be made. Otherwise,	You can easily purchase "goo gone" at the grocery store if you're afraid of adhesive sticking to your windshield.

(Continued)

TABLE 2.1(Cont.)

Item	Paragraph No.	Comment by Reviewer	Resolution by Originator
		modifying this paragraph to be consistent with current practice and policy should be considered.	
10	6.4	This section should also include provisions for POVs to be free of leaks and drips of lubricating, fuel, and coolant leaks. After notification, vehicle should not be brought back on Post until documented repairs have been made.	Requirement added. Additionally, you have now been appointed as the parking lot vehicle inspector to ensure compliance with the new requirement.

I'll admit that those exercises were not a good use of my time or even my best work, but they kept me sane. I'd share others, but that would risk giving away the time and place and I can't afford the kind of lawyer that would require. The funniest thing about the whole saga was that Francis and I never met face to face throughout the entire process. I didn't even know what he looked like until months later …

When construction of the plant was nearing completion, much of my work shifted from writing to reviewing. I was sent out to validate startup and operations procedures as the new equipment was put into service. Many of the systems and facilities were the first of their kind, so the task lent quite a bit of perspective to the more theoretical work that I had been doing while in writing mode. Validation usually went much smoother than the approval process.

One day, a fellow safety specialist named Mike and I were tasked with reviewing the startup procedure for the site's massive backup generators. It was a particularly cold day, so the group had been a bit anxious to get on with it. Typically, a walk through such as this could be expedited if all of the parties involved agreed to keep the chatter and menial comments to a minimum. Typically.

On this day, however, what should have been a formality became a dog and pony show as one of THE FOUR decided to attend. No one would have actually known who the strange new visitor was except for the fact that the plant, which would eventually employ over 1,500 people, had a requirement that all personnel in the field have their names clearly marked on their hard hats. I saw his name before he realized who I was. But it didn't take long before Francis realized that the annoying upstart of a safety technician who had been a constant thorn in his side was now standing mere feet away from him.

I could feel his seething judgment throughout the meeting. The minutes ticked by agonizingly slow as everyone (me more than anyone) braced for what we knew was inevitable. If he remained true to his reputation, Francis would lavish the opportunity to throw a last-minute hand grenade into what had actually been a well prepared and thought out presentation. One could only imagine what kind of off the wall "contribution" he might make live in person.

But as the leader of the operation finished detailing his plan and opened the floor up for questions or concerns, everyone fell silent. He went around the small semi-circle one person at a time to get the required nod of approval. One by one everyone gave him the "I'm good" until he reached Francis. The crowd simultaneously drew in a deep breath and waited. Then … nothing. He gave the OK as well.

Seeing his opportunity at an uncontested green light, the leader dismissed the group. I gave a quick sigh of relief, glanced warily at Mike, and then turned and began to walk away. Mike knew full well why I wanted out of there and followed my lead. We were at least 30 yards away when we heard Francis yell at the top of his lungs, "Hey safety guys! You missed a hazard!"

Mike and I stopped and turned around, mostly in disbelief. Francis was standing at the base of the concrete pad where one of the generators sat, waving his hands above his head. Had I not been there to see the spectacle, I wouldn't have believed it. Without saying a word, we made our way back to where he stood, trying to spot the "hazard" before he had to point it out to us and make us look foolish. I was at a loss.

Neither of us spoke as we stopped just feet from where Francis stood. I raised an eyebrow and gave a questioning shrug, but I wasn't about to give him any satisfaction by appearing concerned so I remained silent. Francis grinned as if he'd caught me dead to rights. This was his moment, his time to prove once and for all that he was superior not only in intellect, but also in his ability to assess safety hazards.

"There's a nail. It could cut someone!" He pointed down to a small wooden step that sat butted up to the structure below the door to the internal compartment. It was a stand-in for the permanent steps that had not yet been installed. A single 16-penny construction nail had been driven into the top board and had penetrated the outer layer of the board beneath. It lay flat against the board it had been intended to be sunken into, pointing straight at the ground. I looked at Francis in disbelief as his eyes glimmered with all the pious arrogance I had imagined they would. I glanced at Mike who gave a shrug of his own, but with more uncertainty than mine. The remainder of the crowd was frozen watching the spectacle, and all eyes were trained on me. I looked once more at Francis as he eagerly anticipated my surrender and for a second I considered letting him win just to make him go away.

Instead, I looked down at the small wooden box once more and gritted my teeth as I formulated my defiance. Slowly, deliberately, I picked up the step, turned it 180 degrees, and then set it back down. As I stood back up I looked at the gobsmacked Francis and shrugged once more, never uttering a word. Then Mike and I turned and walked away a second time, fighting the urge to high-five.

I simultaneously learned and taught a valuable lesson that day. Though it was done in complete silence, everyone in attendance learned the difference between something that matters and something that isn't worth waiving your arms in distress and shouting at the top of your lungs over. That event took me one step closer to discovering how we should view the world when it comes to safety. It would still take a few more years to put the thoughts and theories together, but eventually I would come to live by the idea that success means you have to stop focusing on the trivial and do what matters.

3

BACKSTABBERS WITH ...
"PASSION" FOR SAFETY

We'll dive into the things that matter in Part II but in the interest of starting with a clean slate ... let's look into a few dearly beloved safety traditions that don't. In the past 50 or so years since the safety profession has seen its meteoric rise to a place of prominence in the workforce, the safety professional has been plagued by a rearward facing mentality. We measure negatives (here I go on my rate rant again), we preach "paying attention" to your surroundings, and we squirt compliance checklists out of our cluttered offices like a pack of feral cats heaving up hairballs. We have become speedbumps to progress rather than resources who can teach, inspire, and partner to get s#!* done.

As a consequence of this misdirection, it has become common, sometimes even expected, practice to pompously beat our chests and proclaim from the hilltops that "ALL ACCIDENTS ARE PREVENTABLE." Prominent scientific(ish) minds have taken this mantra and struck gold by selling the idea that 80% (sometimes more depending on who's talking) of accidents are the result of a choice to behave in a manner that puts oneself "at-risk," while at the same time asserting from the opposite corner of their greedy mouths that "nobody wants to get hurt." Think about that for a second.

The very nature of this discussion will lead you down a paradoxical rabbit hole that only serves to bolster the ego of those trying to justify their rocking chair safety programs, but we need to get it out into the open. The longer we pontificate about how educated and brilliant we are, the longer the workers we support get no actual support. I'm not going to ramble on about this forever, but the point needs to be made. We need to stop.

Don't mistake what I'm saying here. I'm not asserting that all safety pros are blundering boobs who can't figure out which side of the sandwich to put the bread on. I *am* saying that the vast majority of us get wrapped up in things that *just do not matter*. I like the way psychologist Clive Lloyd put it in an online debate not too long ago. Speaking about the efficacy of Behavior-Based Safety (BBS) models, he said:

> The pro and anti-BBS "debate" is tired, and not just a little point-less. There have been gains made (and losses too). Its losses does not mean it can't be useful, and its gains don't mean BBS is best practice – it (quite simply) is not.

The subject of the debate isn't what I'm keying in on here, so don't assume I'm diving headlong into the Behavior Based vs. Human and Organizational Performance debate. His point, which I believe was entirely missed by the other parties involved, was that the debate itself is part of a glaring problem with those of us who claim to be working to improve worker health and safety. We consistently, sometimes indignantly, aim toward the wrong targets and measure the wrong things without ever empirically proving we have any effect at all. Yet we continue on the same dead-end paths in spite of never truly moving the dial in a meaningful direction. That's called insanity. Consider the following safety staples:

- We assert that an injury "should never have happened."
- We declare our organization "world class" because our rate is lower than average.
- We cite an employee's "choice" to behave "at risk" when they are injured.
- We cling to tired, debunked "science" and continue wasting both time and money with pre-shift stretching (McGowan, 46–47).
- We write a new compliance policy any time something goes wrong just to make sure we're covered legally.
- We proclaim that Zero Injuries is the only "acceptable" goal.
- We attempt to engineer mistakes out of humans rather than design our system to account for them.
- We dive headlong into the next flavor of the week "safety incentive" program.

- We passionately rally for funding to avoid OSHA fines instead of fixing real problems.
- We spend hours, days, months, sometimes even years analyzing all of the things people did wrong that attributed to their life-altering injury, but we avoid actually taking action to keep those things from ever happening again.

We *should* be looking for solutions that actually make a difference to our workers. We should think about the way our outlook and definition of the word "safe" makes people devalue our goal. We should realize that all of the debates, all of the non-scientific assertions about behavior, all of the write-ups for violated rules, and so on only distance us from the people we have been charged with supporting. Every time a safety professional engages in these types of exchanges, he or she perpetuates the perception that safety is just an extra, burdensome task employees *have* to do which is impeding the other things they *should* be doing.

In reality, it is so much simpler than that. Safety is just *part* of what they do. It has to be, and indeed it already is. If that wasn't true, our lives would be dramatically shorter. The role of the safety professional should simply be to help make that *part* (the safe part) of what they do better, faster, and stronger.

Now, I realize I've gotten a little passionate in the last few paragraphs, so let me tell a couple more fun tales to help explain what I'm advocating here.

Having finished my long slog through the political nightmare where my nemesis Francis still toiled in his embarrassing defeat, I found myself charged with providing safety oversight for not one, but 16 facilities in the mountains of Northern California. I had earned what I considered the "role of a lifetime" (one of several as I would soon learn) and was now referred to as a "Regional Safety Manager" (I would write that in gold sparkles, but you get my drift). It was my first management role and I was bound and determined to make my mark. I still had Nick as my lifeline, even as his health had begun to decline. Had I known then that I would only see him two more times before he passed, I may have leaned on him more heavily at that time, but I digress. I was going to make a difference.

Barb was one of our "master mechanics." And I'm not saying that with any hint of sarcasm. She was a piece of work to be sure. Gruff as any 30-year worker you can imagine, but she was incredibly talented. Any negative thought about her, however, was easily negated when she demonstrated her knowledge and skill. She could troubleshoot and repair any mechanical system in any of our 16 plants with ease. I truly respected her ability to navigate a career which was traditionally "manly" and rival, even better, the most skilled male mechanic on any of the crews in the region. She had earned her stripes through ability and tenacity alone. I'm no social justice warrior, but she was a true credit to women in the industry.

But Barb had one enormous chip on her shoulder. She had been … injured at work. One of her primary responsibilities was maintaining the miles and miles of steam piping the company had strewn about the countryside. These pipelines were the lifeline of the business we supported and required constant monitoring and upkeep. Some of them had been in service for decades, and many degraded at a rate far faster than you might imagine. As those who have worked on pipelines know, nature is a powerful thing. The steam we extracted was no exception, and it was extremely corrosive. Our mechanics were constantly chasing leaks. Often that meant they would need to replace entire sections of pipe.

The funny thing about these pipelines was that many of them were relics, remnants of the "Wild West" that had been put in place long before even our veteran mechanics set out to tame them. There were stories of massive undertakings to wrench 40-inch piping into place against the curve of a mountainside using three or four loaders to pull, while the last of the "cowboy" welders bravely melted their quarry into submission. Bearing that in mind you can imagine there were some incredible forces at play any time a line needed to be broken or cut to be repaired.

I personally saw steam lines being let loose that had been ridden like rodeo broncs even after they were secured and lashed down to account for the stored energy left within. The problem was that they were completely unpredictable. You might, for example, strap down a pipe in anticipation of it bucking upward, and instead it would buck to the right.

Barb was unfortunate enough to feel the wrath of that pipe one day. She was unbolting a seemingly innocuous flange in order to perform some routine repairs. As she removed the bolts, there did not seem to be any significant pull or flex in any particular direction. She removed one bolt after another with ease. Then, only one bolt remained. She began wrenching on it with ease and suspected nothing as it did not appear to be under any undue strain. The bolt backed out smoothly and then ... snap! The 40-inch pipeline sprung loose wildly, sending Barb and her wrench reeling backward. The large wrench contacted her face with force, knocking out her four front upper teeth.

"She shouldn't have put herself in the line of fire!"

"That accident was completely preventable."

"If she had planned the job properly, that never would have happened."

Do those statements sound familiar? I imagine they do. We have made a religion out of second-guessing the actions of employees who were in the wrong place at the wrong time. But the truth is that it wouldn't make any difference if Barb's injury *had* been preventable. Would you like to know why?

BECAUSE IT ACTUALLY FREAKING HAPPENED! I dare you to tell Barb to her face that she should have done *this* or *that* differently. You'll likely experience her fate if you do (as in, she would knock *your* teeth out). To her, at that moment, there was no alternative. There was no indication that a hazard existed, so why would she have behaved differently? Did she *choose* to smack herself in the face? Of course not. So why do we "professionals" hang our hat on our armchair quarterback analytical skills by stating that an injury was preventable, that it shouldn't have happened? Even if an event *might have* been prevented under a different set of circumstances (an argument that could never be proved), the reality of what really occurred can never be changed. Arguing the point only makes you look like that sandwich boob I mentioned earlier. Safety professionals who fail to see the underlining arrogance that accompanies statements such as those mentioned above do no service to themselves or the profession.

I understand that this activity is done in the vein of "learning," but ultimately it comes off as condescending and detached. It serves as yet another unnecessary chasm between the safety pro and the worker. A glaring example of our inability to understand what they are going through and thus a huge missed opportunity to help make things better for the next person tasked with loosening the bolt on that spring-loaded pipeline.

So, I challenge you to take this to your workplace. The next time someone starts talking about how a person exhibited "at-risk" behavior or should "never have been injured," ask them if they would honestly have done anything differently if they had been in that moment. Any intellectually honest person would have to answer, at the very least, "I don't know," if for no other reason than the fact that the worker didn't know the outcome before it happened. Retrospective understanding of an event can only come from hindsight. There are plenty of people who would argue that point, but that's precisely the problem.

So why do we see fit to judge the past based on what we only now know in retrospect? I would submit that any time we focus on the consequence of an event it's a poor reflection of our ability as "professionals." The consequence (good or bad) is irrelevant because it cannot be changed. We do it because it's easy, though. That is why companies judge safety performance based on rates. It is a path of much less resistance to judging the past than forging a new path into the future. Put another way, it is easier to fixate on the consequence of an incident (the lost workdays, the OSHA recordability, etc.) or find someone at fault than it is to forge a new, risky path toward true improvement.

The next example might seem a bit off topic and even bitter (you'll see why at the end), but I firmly believe it is the small things that make a difference and often those small things drive our cultures without us even realizing it. I tell this story because it's a prime example of how easy it is to get off the rails and focus on the wrong things. On *this* particular day, I sat in my office in blissful ignorance when one of the plant managers came in to ask a safety question. It was innocuous. Innocent even; and we actually had a really good conversation about it at that moment.

This story is set in the midst of turmoil following some extremely poor management decisions revolving around safety (more on that later); and everyone was hypersensitive about it. That being the case, the manager, who just so happened to be new in his role as well, came to me asking for some clarification about an issue that had happened the night before.

At his plant, one of six large cooling tower fans had suffered a catastrophic failure in the middle of the night. One of the fan's blades (which turned out to be 14 years past its expected lifespan!) had broken apart at its coupling while running at full speed and had shattered in two other places, inertia sending the pieces slamming into the fiberglass shroud which guarded the fan housing. At the tremendous speed it had traveled, the broken tip of the blade had punctured the shroud and embedded itself at about five feet above the walking deck. The shroud was concave around its centerline making it roughly two feet wider at the top and bottom. The blade had stuck flush against the outside of the shroud just about dead center of that concave section (kind of like Francis's nail come to think of it). To give you a better idea about what I'm describing, I could stand with my feet touching the bottom of the shroud and I had to reach out those two feet in front of me to touch the damaged section.

The question asked was one you might be anticipating. Was this incident a "near miss?" Many would probably have answered it differently, but remember, we're discussing things that don't matter here. My answer was no. In fact, my exact words were "No, but that's a pretty serious safety issue and we definitely need to make sure it doesn't happen again." My reasoning in saying no, of course, was that no one was anywhere near the fan blade or even planning to be (it happened in the middle of the night). No one was "almost" injured, but something catastrophic absolutely did happen, there was nothing "near" about that. Had an operator been standing near the shroud my answer would have been different. At the time he agreed with my reasoning and also committed to making sure it didn't happen again. He then left and I felt pretty good about the conversation.

Fast forward a week or so and we were both sitting in the monthly safety committee meeting for the region. I was reading through the list of incidents. Reading from my notes, I reported that there had been no near misses during the period. And that's when the tables turned. One of the environmental technicians whose job was to take periodic emissions readings on the cooling towers threw up his hands, turned beet red, and started yelling about how that was BS. His life was in danger and it *was* a near miss. And then, my ally the plant manager said, "Yeah, I told Jason it was a near miss, but he didn't want to do an incident report on it."

I shook my head and glared at him. Then the meeting devolved into a non-productive, emotion-driven soapbox consumed by a debate about semantics. The fan blade certainly was a serious issue. Whether it should have been classified a near miss or an equipment failure should never have played a part in the conversation. But because we were so stuck on definitions, we missed an opportunity to learn, grow, and make the situation better. If I had to bet on the condition of the fan blades at that plant even now, I would guess they're still past their expiration date just waiting to come apart again. Who knows, maybe next time it *will* be a near miss ... or worse.

4

RISK IN REAL TIME

I've been accused of many things throughout my career, mostly false. It's happened so often that I sometimes equate being a "safety guy" with used car sales (I did it, so I speak from experience … my wife will tell you I'm not good at sales though). I'm sure many of you have been met with similar disdain and dislike just based on your position. I've dealt with threats of violence, accusations of ignoring employee concerns, and even giving other employees cancer (yes, you read that correctly … I am *that* powerful … apparently). Rarely, however, I'm accused … no wrong word choice … I'm lauded for having wisdom beyond my years. This occurred recently with my current boss, a vice president of the company.

My statement to him was a simple one, but I truly believe it's the beginning of a paradigm shift in the way we should evaluate our workplace safety performance. It's something tangible that we can begin to use in order to look at the past and glean meaningful preventive measures. The statement was this: "People don't assess risk in real time." He took his sweet time evaluating what I had said, tempting me to doubt my thought as the seconds ticked by. But to my surprise, his response was to tell me that it was a very wise observation. In context, that doesn't happen often when I speak to someone 18 years older than I am.

I'm no behavioral psychologist or neuroscience expert, but this theory is based on 15 years of observation about how we react to risk within our environment. Everyone encounters risk. Maybe that risk is your commute to work, or maybe it's simply the way you squat down to tie your shoe in a position that's compromising to your body's natural strength curve or neutral position. Or maybe the risk is potentially life-threatening and requires detailed analysis. No matter the point on that scale on which your particular risk exists, you only have three options for handling it. Three. No more, no less. None of them is foolproof, so keep that in mind as we work through this.

I've been a serious recreational weightlifter for years, so I'm going to use that experience as my case study for this exercise. Take, for example, an elite level powerlifter who is training for a meet in which he's attempting to set a regional record in the deadlift. Aside from the hard work, dedication, and drive just to put in the training, his best option for making that goal a reality is the first (and best) way to address the inherent risk of lifting say … 805 pounds. When the moment comes he has mapped out every queue, every breath, and visualized the outcome thousands of times before the attempt is ever made.

That's your *first option*: Plan your work. Plan every detail. Analyze every potential point of failure and draw a "map" around those hazards that will provide your best, most likely chance of success. When that is done, you step to the bar, position your feet mid-way under it, tighten your lats, reach down to the bar, pull the slack out, engage your glutes/hams/quads, and attack! From that point, you'll either succeed or fail, but the task is all skill, training, and muscle memory from that point on.

Let's not forget that at this point our lifter is using every ounce of himself to impose his will against 805 pounds. There is no arguing that there is a huge inherent risk in that task. So, let's assume now that he's moved the bar three inches off the platform but the weight is too much on this day. The planning is over. Now he is only able to react. That is your *second option* when dealing with risk: Reacting. From this point, he might plow through and grind out the rep … He might feel fear creep in and drop the bar … He might snap under the pressure and experience a terrible injury. The point is that reacting to risk is a crapshoot. Your odds may vary depending on your skill and experience, but it's still a crapshoot.

Since we're talking safety here, let's say he snaps himself up and tears a bicep tendon (a common injury associated with the deadlift). The bar drops and so does the lifter, curled up in pain. At this point, there is only *a third and final option* available to address the risk. That is to analyze it in hindsight. This is where we safety professionals get ourselves in trouble.

Hindsight is a powerful tool, but it's also a drug that makes us feel more powerful than we really are. With hindsight, we can pinpoint the exact moment when that one muscle gave out or identify that the lifter's breathing wasn't quite right. The key is remembering that you did not have that knowledge prior to the event. Competitive sports teams understand hindsight in a way most safety professionals have forgotten or even refuse to accept. They scrutinize game day footage second by second, but not simply to dwell on the past and wallow in their failure (or victory for that matter). They watch every game, win or lose, and they use that information to improve. The outcomes (consequences) of the past have no value in that learning moment because focusing on the win or the loss puts blinders on that shroud your vision from the key lessons you need to learn in order to perform better next time.

For some reason, though, safety practitioners hang their hats on retrospect. You've likely heard a million reasons justifying why we do it, and as a learning technique there is nothing inherently wrong with analyzing what happened in the past. Where it goes off the rails is the moment when we start proclaiming that an incident (which has already occurred) was 100% preventable. Some even try to provide a scalable list of "preventability." I've already kicked this dead horse a couple times in the previous chapter, but I think it's worth really driving the point home (mainly because it just plain annoys me that this is a "staple" of our business).

What kind of arrogance does it require to sit back in a padded office chair wearing a clean, crisp polo and judge the way *someone else* reacted to a risk? If that polo-wearing know-it-all can't recognize that it was partly his failure for missing an opportunity to identify, plan, then train to reduce the impact that risk could have on an employee, there really is no hope for him. We have to stop telling people past accidents were preventable. It is true that the future is not written, and I think that's probably why this subject gets so muddy.

The "fix" for this downfall is to separate the events of the past from the plans you make to prevent future failure. Separate them deliberately and definitively. When someone reports an incident or injury, tell them thank you for bringing it to your attention and move on (assuming they're not bleeding out on the floor). Then figure out how the system broke and exposed that worker to too much risk. If any solitary change in the way we "manage" workplace safety would have a larger impact on the profession, I've yet to discover it.

One final thought on this topic. Why is it that we often only focus on safety "incidents" because they resulted in an injury? Injuries are actually the smallest minority of consequences that occur as a result of the work our people do. Why wouldn't we instead actively, passionately focus on the opportunities we have to plan rather than react? If we did that, we wouldn't have near as much need to armchair quarterback in retrospect. And maybe, just maybe, our people would see us putting in the work and they might be inclined to join in on the effort.

5

INSPECTOR BOB

Around the same time as the fan blade incident, I received an unexpected call around 10 a.m. one morning. I remember only having been a regional safety manager for about three months at that point and had not even made a full sweep of my region and visited all of the 16 facilities. The call was one that often fills safety people with dread: OSHA was on site to investigate an employee complaint. I was confident, however. This was *not* my first rodeo (don't ever believe that by the way).

I quickly gathered up my gear, along with some spare items the inspector might not have brought along. The "region" was fairly small, but navigation was not an easy task considering all of the narrow, winding roads. It took me a good 25 minutes to make it to the guard shack to meet the man I now refer to as "Inspector Bob." It was clear he was not impressed with my level of urgency. I handed him my business card and reached out to shake hands as I introduced myself.

"I'm Jason Maldonado, Regional Safety Manager," I said. He took my card, but shrugged off the handshake and simply flashed his credentials as is required. I tried my best not to take offense, but the denial felt deliberate and calculated.

"So … you're basically a 'safety coordinator' then," Bob replied condescendingly. I'm positive my jaw gaped a bit after that comment, but I stood my ground.

"No," I said pointing to the card I had just handed him. "I'm the Regional Safety Manager. We actually have 16 facilities throughout the mountains here and I'm responsible for safety oversight at all of them." Our exchange was not off to a good start. "What brings you up here today, Mr. Inspector?"

"I'm here to investigate a complaint at the Boulder plant," he said. (I've changed the name of the plant for obvious reasons.)

"OK, then. Let's head down there. What would you like to look at?" He looked down his nose at me after that question.

"Oh, I can't tell you that. If I did, you'd know who made the complaint and you would retaliate against them." I was taken aback in disbelief. He didn't say I "might" retaliate; he had already passed judgment. It was clear he viewed me as the enemy. An evil representative of a nefarious corporation which aimed to make its employees suffer and die slow, agonizing deaths. I'm sure my complexion was getting more rosy (an unfortunate tell of mine) by this point, but I tried once again to hold my ground.

"Actually, Mr. Inspector, that's not possible. I've only been with the company for three months and I don't know any of the employees who work at that plant. I have no problem taking you directly to the area of concern." He was unwavering though.

"No, you would retaliate!"

From that point on, I resolved to play unrelenting defense. It wasn't what I wanted to do, but Inspector Bob was unlike anyone I have ever met, even to this day. I motioned to our vehicles and asked him to follow me to the plant. I drove with the radio off and stewed in silence during the additional 30-minute drive.

When we arrived, I meticulously walked Inspector Bob through the 30-page visitor orientation manual that was typically used for contractor training (a half-day event, mind you). I dug into every detail and made the "lessons" tick by as slowly as I could. Partly as a Cover Your Ass, partly out of spite for his attitude, and partly because I needed time to figure out what he was after. He wasn't nearly as smart as he liked to believe (and undoubtedly still does), so I figured out pretty quickly that he was there to investigate an alleged chemical exposure. That prompted me to step away briefly and call in the company's industrial hygienist for backup. But he was over two hours away. Thankfully Bob was in no hurry. He wanted to take me on a wild goose chase as soon as he noticed that I was trying to sniff out his agenda, hoping he could throw me off the trail.

When I finished his safety orientation, I tried one more time to get him to give up his objective. He didn't bite and suggested we "just take a walk." Every alarm in my head was going off by this point. On the list of things I wanted to do that day, giving free rein to this type of "gotcha" inspector, was about three levels below a prostate exam. I began counting down seconds until my backup arrived.

First, we meandered through the rundown women's locker room. There were no women assigned to the plant. You could say housekeeping was less than stellar in that area. Nonetheless, Inspector Bob snatched up a roll of toilet paper and tore off one sheet at a time every time he encountered a ventilation fan. He explained that he was so good at industrial hygiene that he didn't need instruments to tell him what met the standard and what didn't. He simply needed to verify that the sheets of paper stuck to the vents. The locker room proved fruitless, though, so we moved on.

As we exited, a mechanic named Seth walked past us, noticed something was off, and tried to shuffle away before getting sucked into the mess. I actually knew him since mechanics roved from facility to facility. In retrospect, I'm sure that confirmed to Bob that I had been lying about my knowledge of the plant employees. Bob caught onto Seth's angst and glared at me.

"I'd like to speak to him," he said.

I sighed and then caught Seth's attention with a quick wave. He reluctantly walked over and I provided introductions.

"Seth, this is Inspector Bob. He's from OSHA and he would like to speak with you. Bob, this is Seth. He's one of our mechanics." Bob said nothing in reply, opting instead to flick his wrist at me in a dismissive gesture. I sighed again and walked out of earshot. I could see the disbelief in Seth's eyes as I moved away. Even after already having spent over an hour with this gem of a man, I was still a little surprised by his outright rudeness.

From that point, the "inspection" became ludicrous. He didn't get what he was looking for from Seth so he asked us to step into the administrative offices. I knew his quarry was nowhere near that area of the plant, but obliged anyway. As soon as we stepped into the door he asked a question that has forever changed my view of humanity in ways I never would have imagined. The door of the office area slowly creaked shut as Bob fixed his gaze on a drop ceiling tile with a small brown water stain on one of its corners. Look up at your office ceiling the next time you're there and you'll likely find a similar sight (incorrect or not, it's pretty common).

Bob pointed his crooked finger up toward the spot. "What's this indication of mold I see on this ceiling tile?" I looked at him and cycled through responses as he shifted his body awkwardly and then took two exaggerated, deliberate steps to his left. As I opened my mouth to answer, he cut me off with a cartoonish "gee whiz" type gesture and then in a screechy, cartoonish voice he answered his own question: "I don't know Mr. Inspector. Maybe this company doesn't care about taking care of leaks that could create harmful contaminants." I shook my head in disbelief, half in denial about what I had just witnessed. That was only the first of many such episodes.

We slogged on. He looked at every corner of every room and every piece of equipment at that plant. We walked for hours, "inspecting" a plant that sat on just two acres of land. "Alter-Bob" liked to talk *a lot*, so I had figured out what he was after shortly after he began answering his own questions. I also knew he was going to be disappointed when we finally got around to it.

Over the previous 30 years, the company had done extensive annual analysis of the natural steam used in our process. It was a well-known fact that there were trace amounts of arsenic in it, but at nearly undetectable levels. The only place it ever appeared in quantifiable amounts was ... believe it or not, at the notorious cooling towers. At well past the four-hour mark, Bob and I were standing at the base of this plant's cooling tower which only consisted of four fan compartments. I had outright volunteered to show him our arsenic sampling data, but he was the most indignant person I've ever met. He refused to admit a "coordinator" like me had him figured out, so the goose chase continued. Then he asked me one of the first non-rhetorical questions of the day.

"Tell me, Jason," he said. "Does it rain here?" I knew what he was after, but I wanted him to earn it.

"Yes Bob. It does." I pointed to the sky. "See those white fluffy things up there? When they get upset, they cry and their tears drip down onto the ground. We call that rain." I know that response may seem exaggerated, but that's what I actually said. I'd be willing to bet most of you would say that (or worse) after having endured this man. But Bob was unphased.

"No, no. Does it rain ... *here?*" He motioned at the area around the cooling tower.

"Yes, Bob. That happens too. It's called cooling tower fall-out. Would you like to see our sampling data regarding what we've found in that material now?" He didn't have a chance to answer because at that moment our industrial hygienist showed up. From that point on, I was chopped liver.

The day's investigation concluded shortly after, but I was stuck for the next six months answering Inspector Bob's "dis-covery" questions. As a side note, if you ever have the pleasure of dealing with an inspector who has multiple personalities, answer all of their closed questions with a "yes" or a "no." It makes them exceedingly happy. He never found what he was looking for and we eventually learned that the "complaint" had come from an employee who had been fired months earl-ier for smoking pot. As revenge, Inspector Bob cited us for an administrative error in our site safety manual and the saga was laid to rest.

The Inspector Bob escapade is a story I enjoy telling in person maybe even more than I've enjoyed writing it. It's a comedy act that I didn't even need to make up. But it's also perplexing to me if for no other reason than to ask that resounding question that has haunted me throughout my entire career. How did any of it, the dog and pony show, the hours of digging, the research and legal fees, the employee interviews, any of that ... How did it make anyone safer? For that matter, how much of the time does your average safety professional actually contribute to improving worker safety through any of the stereotypical programs we "manage"?

You and I both know the answer to that question. If you've ever asked it and wanted a real answer, I'm going to ask you to take a chance with me. Let's do some things no one has ever done before and build a framework for safety that works. It will take faith, fortitude, and likely garner more than a little criti-cism from the very people who claim to work toward the same goal, but never waiver from their tired mindset. If you're will-ing, I'm ready. Lets go!

PART II

FOCUS ON WHAT MATTERS

Writing this book thus far has been as much a journey of self-discovery as it has been one of reflection. My hope is that these bits of knowledge and tips designed to help people get better are not taken as me sitting on some high horse in a lofty corporate office, basking in my prestigious accomplishment. I am your average "Safety Professional" who has some fancy letters at the end of his name just like thousands who have come before me. Those letters don't make me special. At one point I actually considered calling this "Memoirs of a Mid-Level Nobody." While that's definitely in line with my dark, sarcasm-based humor, it isn't really indicative of what I'm aiming to do here.

The stories in this book are moments in my time. They're not chronological, and when taken one at a time, they don't make any monumental statements. Some are certainly powerful, and I hope I've done them (and the people who taught the lessons) justice. I'm no superstar, but I have seen real success in safety management, if even in mere glimpses or momentary flashes. Most of these lessons were learned the hard way, and as I've observed over the years, there really aren't many resources for someone who wants to make a difference to learn how to accomplish that. There are libraries full of safety books that will teach you how to interpret regs, or audit worksites, but I've yet to find anything that can provide a path to real interpersonal success in this field. This is where that rubber meets the road.

In Part I I did not even attempt to provide an exhaustive list of all the pitfalls safety pros subject themselves to. That's not a task for one person. You likely have experienced and witnessed your own examples and I'd be glad to share those with you on social media. Provide your experiences and include the hashtag **#relentlesssafety** along with it. Or you can leave a comment on relentlesssafety.com. Maybe we'll learn something from each other.

This part is a new direction though. I'm going to keep telling my stories because I think they're fun and, quite simply, that's how I learned these lessons. But the game is changing from here on. One easy criticism of those who speak out about the problems with the status quo in any facet of our society is that their musings usually amount to nothing more than internet rants about what's wrong, never providing any meaningful solution or alternative. Well, you're in luck. I have *all* of the answers to our safety culture conundrums!

OK, that's just not true. No one has every answer. But we all have some. In the following chapters, you'll relive some of the most impactful episodes of my career. Along with each I'll provide a tool that you can take directly to your workplace and use to begin making some immediate, substantial change toward becoming a better safety professional. That's not arrogance speaking either. If you do these things, you *will* make a difference.

Sound too good to be true? Well, that's because claims like the one I just made are usually associated with something "easy." These tools aren't easy. If you use them, you're going to have your work cut out for you.

6

PADZILLA

Stepping out and putting yourself on the line for the pursuit of something better or even just something *more right* is never easy. Sometimes the risk doesn't pay off. Sometimes the person who takes the chance never sees the result. But the optimist in me says it's always a chance worth taking.

I literally spent hours today speaking to another safety "professional" about the very ideas I've been covering in these pages and I can't help but feel disgusted with the way the conversation ended. Not specifically because I was right or he was wrong (I'll let you read between the lines regarding my thoughts on that issue), but more because our disagreements are proof that the chasm between real change and the status quo is miles upon miles wide. But, just as I told him, nothing will change until we change. You can fold your arms and disagree with that truth (he did), but it won't make it any less true.

The experience reminded me of a powerful lesson I learned toward the end of a one-year tour in South Korea while I served in the US Air Force. It was my first duty assignment and it was filled with many lessons and good times. I actually spent a total of 13 months at Osan Air Base because I had traded a stateside assignment with another young airman who wanted to stay close to home. He had graduated technical training earlier than I had, so my ship out date had been moved up. I spent the first few days on base in near solitude, accepting that the closely knit group of guys I had been training with in tech school were scattered elsewhere throughout the world. Then, to my surprise, "Padzilla" showed up.

His first name is Jason as well, so neither of us ever referred to each other by that moniker. His nickname, as most in the military including mine, was a play on his last name (and actually something he didn't earn until midway through the year). Korea was a party from your first weekend. Initiation meant hitting every bar on the small strip in the town of Song-tan until you forgot you were in a new country. Guys that had been there longer called it your "green bean" tour. The best part was that the legal drinking age was only 20. Padzilla, who was 19 years, 11 months old, figured he was close enough and joined in as soon as his jet lag subsided.

While I probably would have shared his logic, it was no concern of mine. I had just had my 22nd birthday and considered myself a well-qualified, legal drinker at that point. Regardless, the bartenders in Songtan couldn't care less about how old anyone was. If you were a "GI," you were legit. No one was ever carded at a downtown bar and the younger guys were in hog heaven. No cover stories needed, no fake ID required. Just walk in, order an "OB," and a shot of Soju and you were part of the party. Padzilla fell into that trap quickly, but he hadn't learned the one required trick: don't go home if you go underage drinking.

It was true that the bars didn't care who or how old you were, but we had a curfew most nights. It wasn't a hard thing to follow but the rule was that you needed to be "in" by midnight on weekdays and 1 a.m. on weekends. All that meant was that you had to be back on base, checked into a hotel, or staying with one of the few service members who lived off base (which rarely happened). Padzilla fell in line on his first weekend. Given our general low-fund bank account statuses in those days a hotel was out of the question, and he figured there was no harm heading back to base after a night on the town. And there wasn't … if you were 20. So, on his first weekend, in this new, foreign land, Padzilla let loose.

To be fair, I wasn't with him the night it happened (and I'm not even sure it actually was his first weekend), but I saw more than a few people try to skirt under the wire unsuccessfully. The issue was that getting back on base was not as simple as waving at Joe the security guard and then walking the mile of shame back to your dorm to sleep off the night's bender. Security Forces manned our base gates and they relished the power they had over the drunks lined up like cattle to be granted access back to sovereign US soil. Most of them were decent people, but some just wanted you to give them an excuse.

In Korea, if you were drunk (even a little) and you had an "incident" it meant you were getting thrown in the "drunk tank" for an indeterminate amount of time and then subject to non-judicial punishment. Depending on the severity of your disorderly conduct, that could mean anything from a verbal reprimand, to formal court martial proceedings. Padzilla caught a bad break on his first night out and had his ID checked by the one bitter airman who was jealous that he had been stuck on guard duty all night rather than partying with the rest of us. One month shy of his legal, 20th birthday Padzilla got clipped.

Our squadron chief was a fat, insecure, 29-year near-retiree who loved making examples of juvenile delinquents. In fact, his "welcome speech" to all airmen entering his squadron included some form of the phrase "before this year's over, you'll be in my office getting an Article-15 for something stupid." I remember when he said it to me and the other new airman who had arrived on the same day. As you might imagine I couldn't help but respond by telling him he was wrong.

"You don't know anything about me. I won't be back here getting in trouble for anything," I said. I did end up in his office for something else. He had rolled his eyes at me at the time, but of course I was right. I made it through that year unscathed. When I did end up in his office it was because I had been selected for an early promotion program shortly before my tour ended and my prospect of success looked good (spoiler alert: I got promoted early). Padzilla had a much tougher go of it, though.

He ended up being an excellent "troop" and a very talented technician. He was also an ace at driving a forklift on two wheels. Along with a small group of other guys from our shop we watched each other's backs, but he was always trying to crawl out of the hole that one bad weekend had put him in. He hadn't been given an Article-15 (the military's first step for kicking someone out on an "other than honorable" basis) for his underage drinking, but he had received a very pointed Letter of Reprimand (LOR). I imagine he had to work twice as hard as me and some of the others to prove that he wasn't the idiot drunk our chief wanted everyone to think he was.

Fast forward to the end of our tour. As was customary, most who completed their year of hardship were recommended for a commendation and/or meritorious service medal. Not necessarily because we had really achieved true greatness just for spending 12 months in a foreign country, but because "technically" the Korean war never ended and we had served in a war zone for the duration. It is a thing to be proud of for sure, but I would never place myself upon the same pedestal as those who have served in the relentless conflicts in Iraq and Afghanistan since 2001. In my mind, they are the true heroes. As my months dwindled down, my supervisor began preparing my medal package for review and signature by our commander.

I can remember vividly sitting in our "day room" (break room for those who don't understand the lingo) observing, not merely overhearing, a heated conversation between my supervisor and our shop chief, Master Sergeant Brent. The supervisor strutted into the room with a crisp blue folder and a smirk of pride on his face as he approached Sgt. Brent. Brent was a giant of a man (at least in my mind, I'm 5 feet 6 inches) and one hard SOB. He was born and bred in West "BY GOD" Virginia, and damn proud of it. I had seen him perform some incredible physical feats over our year together, but his real strength was his wisdom, authority, and genuine concern for his people.

The supervisor grinned as he held out the folder. "Here's Maldo's [my nickname] commendation package, Sgt. Brent." Brent cracked the folder and took a quick glance. Then looked back at him. The air in the room became icy cold as we all sensed something was about to go down. A couple of us even peeled our attention from the episode of SpongeBob Squarepants that played on the small TV in the corner.

"Where the hell is Padilla's package? They're both due at the same time," said Brent.

The supervisor shriveled a little and took a small gasp of breath. We could hear him gulp as he opened his mouth to respond in little more than a whisper. He obviously knew Padilla could hear everything and I've always assumed the supervisor was trying not to hurt his feelings.

"Padilla has an LOR for underage drinking, sir. There's no chance the commander will sign off on a medal for him."

Brent eyed him up and down and let him stew for what seemed like an eternity. I'm certain no one in that room took a breath for at least a solid minute. "That's BULLSHIT!" He bellowed at the top of his lungs, glorious West Virginia twang piercing the silence. "Just cuz you think something won't happen ain't no reason not to try! Get it done!" With that he threw my folder back at him and walked out of the room.

I was awestruck by that moment. There are so many levels of learning to take from that exchange, but the one that has stuck with me more than any other was the courage Sgt. Brent had to defy the odds and force something that was an improbable possibility. He forced an issue that only one in a thousand would risk sticking their neck out for. To me that was life altering to witness. The icing on the cake is that when it was all said and done the commander signed that medal package, no questions asked.

If you're looking hard enough, I think everyone has opportunities like that in life. Speaking from experience, they are downright terrifying. In the next few chapters, I'm going to challenge anyone of you who has stuck with me this far to start taking some of those risks. I've talked quite a bit about all of the stupid things that we safety "pros" do on a day to day basis, but none of that amounts to anything more than a whiney rant if I can't offer a meaningful alternative. Personally, I can't stand it when someone comes to me with a problem but has not put any thought into how to solve it. So, to that end I'm going to do my best to provide some tools that I know will make a difference. I call them my "Stupid Simple Toolkit."

Again, if you've stuck with me this far, you're willing to accept that the current system is broken and in need of a full 180-degree shift. These are ideas that *will work*, but don't for a second fool yourself into thinking that "stupid simple" equates to "super easy." This will be some of the hardest work any of us who have chosen this field will ever engage in. Commitment is required. Half-ass attempts will fail. In the words of Sgt. Brent, let's get it done!

Stupid Simple Toolkit Item 1: Nut Check

This tool is the least "practical" tool I'll be discussing. It's more of a call to action than a tangible thing, but it's a logical place to start. We get so wrapped up in what we "plan to do," what we're "going to start tomorrow," what would "be really great." You get my drift. Just as a fun bit of fact: this tool is actually something my older brother told me when we were teenagers. It was his go-to trick for picking up girls. I was a shy kid with no self-confidence and didn't even know how to talk to a girl, let alone ask one out. His advice was simple: "Grab your nuts, make sure they're still there, and get to work." I never mastered it back then but I'm married now so I must have gotten it right at least one time.

The point is that the time for thoughts, hopes, dreams, and talk about future plans is done. Now we act. This could mean something as simple as swallowing the lump of trepidation in your throat and calling a spade a spade the next time an executive makes a bad example for safety. It could mean scheduling that training you've put off because getting people to attend is such a hassle. Whatever the task, stop thinking and start doing. If you want to really make an impact, make your first move with TOOLKIT ITEM 2 (I have to do something to keep you on the edge of your seat).

7

AIM AT THE WRONG TARGET, GET THE WRONG SCORE

I'll never have the illusion that my time in the Air Force earned me a spot alongside those who have bled and died for our country. I was assigned to a fairly low-risk field, although its inherent dangers were certainly life and death. Long before any of that was apparent to me however, I found myself as a "trainee" in the 222 Squadron in Sgt. Williams's "Big Dog" flight. The Air Force was in a time of transition in the early 2000s and the country was still reeling from 9/11. I had joined out of desperation, and not just a little lack of cash. I was a 21-year-old college graduate who had a degree in "talking good" (Communications), and no idea what I wanted to do with my life.

My younger brother had joined in late 2002 and saw me struggling to find myself as a singer/songwriter in Nashville who really didn't have the chops or the drive to make that happen. He was as true-blue as you could get at that point and told me to talk to an officer recruiter. I hadn't had any luck finding a decent job since college, but he was confident the Air Force would want my skills ... they couldn't have cared less though.

I was sent away from the officer recruiter just as fast as I had rushed in the door. They were interested in engineers and people with mathematics degrees, not someone with the academic equivalent of having gotten a gold star for putting fancy words in a particular order. I was at rock bottom at that point, however, so I swallowed the half an ounce of pride I had left and went to enlist. The eager staff sergeant who greeted me at the recruitment center almost turned me away once he learned that I had a bachelor's degree, but I convinced him the Air Force was what I wanted to do. In all truth it was because I was beginning to see the seasons of my career for what they actually were: stepping stones.

On some level, even then, I knew that the propaganda about my college degree being worth its weight in gold was only half of a lie. With that thought in the back of my mind, I had resolved to enlist, get my mandatory "five years supervisory experience," and then get out and get the job I deserved. My career did turn out that way, but the truth was that I didn't deserve anything. Just as any wise, successful person knows, you earn what you get. I was, and still am, no different.

In the weeks before I headed off to basic training, I called my brother and asked him what to expect. He'd had a rough go of it in basic because he had been selected as an "element leader" (other branches refer to them as "squad" leaders). He also had a bit of a chip on his shoulder and the added stress of having his first wife in his "sister-flight." His instructors had used that fact to play some less than savory mind games with him to try to break him down. In spite of all that, however, he had made it through and was now in Technical Training School to be an F-16 crew chief. I had even attended his basic training graduation and met his drill instructors.

His answer to my question was simple: "Keep your head down, take the easiest job you can get, and don't volunteer for *anything.*" Then he added one small, highly unlikely caveat. "You won't have any problems unless you get the same squadron I had … but that probably won't happen."

I shipped out from the Military Enlistment Center in February of 2003 and because I had a college degree and was considered the "highest ranking," I was put in charge of eight other "trainees" who were to be my classmates. My group arrived at Lackland Air Force Base in San Antonio, Texas at nearly 2 a.m. and played pick-em up/put-em down with the instructors for what seemed like hours (it's exactly what it sounds like … pick your bag up, then put it down). One humorous moment I'll recall forever from that night was when one drill instructor stepped out, mid-torture, and stopped the sadism long enough to call out the names of the recruits who had been guaranteed a spot in the Air Force band. They were pulled out of formation and shuttled off to a less hostile proving ground, lest their trumpet fingers become strained. At least that's how it seemed.

Once the greeting crew was done with us, we were divided up on buses and sent off to our assigned squadrons. For the life of me, I couldn't remember which one my brother had told me he had been assigned to. In the dead of night, I watched one training dorm after another pass by until finally our bus turned left and parked in a dimly lit alcove under my assigned dorm. I breathed a sigh of relief because I was at least 38% certain I had not gotten dealt the same hand as my brother before me.

Your first night in basic training is designed to disorient and fatigue you. On top of having traveled for an entire day to reach the base, we were now being subject to a non-stop barrage of nonsensical directions, corrections, and downright outrageous demands. I felt like I had the edge because I knew this was just part of the game. The tear you down/build you up design of military training.

As soon as I stepped from the bus I was snatched up, along with three other trainees, and pulled aside from the rest of my flight. The man who pulled me out was short, like me, and sported the quintessential drill instructor's moustache which had been trimmed to perfect regulation specifications. He didn't introduce himself, but the name on his Basic Duty Uniforms (BDUs) seared itself into my mind and began instantly making me question my confidence ... *Robledo*. I had seen that name before, but surely ... there was no possible way?

Staff Sergeant Robledo stood at parade rest and bellowed to the four of us. "You four are my road guards. Beginning tomorrow morning, you will ensure that your flight does not get plowed over by oncoming traffic each and every time they cross a road."

From that point he proceeded to show us how to run out into a road in front of our formation of troops and stop traffic while they crossed safely to the other side. For a moment I was elated. This was one of those "easy" jobs my brother had told me to get and I had scored it without even trying. Once that quick training was complete, we were sent to the day room to join our flight for role call.

In the day room, we were quickly taught to sit "at attention." All that meant was that we sat *Indian style* ... sorry "crisscross, applesauce" with our hands on our knees and backs razor straight. Then our lead drill instructor, Technical Sergeant Williams, took role. It went agonizingly slowly.

"Adams, Adkinson, Beale."

He paused only long enough for a simple raise of the hand and corresponding "Here sir" response.

"Macdonald, Maldo ..." He paused midway through my name and my fate finally registered. Slowly his dark, cold gaze met my eyes without even having to scan the room. "Maldonado?" he asked. "You have a brother come through here about three months ago?"

I gulped. "Yes, sir."

He looked back down, scribbled something on his role sheet and simply said, "Aight." No one else could have known what that exchange had meant, but I knew I was screwed.

The next morning, I was more driven than any other road guard in the history of the military. I would put my "stop signal" palm up in front of a tank even if it meant I'd get flattened into a trainee wafer if the driver failed to stop in time. I was so determined that I would have been willing to peel myself off the pavement and run down the driver of that tank and tell him he needed to put it in reverse and back up until my flight crossed safely. Even after having been run over.

I was exceptional at that job. In fact, I'd be willing to bet my first born (just don't tell him) that no one has ever been a better road guard since. But none of that mattered. Tech. Sgt. Williams already had his eye on me.

He let me bask in ignorance until the early afternoon of our third full day of training. We had been awake nearly the entire time (save for a couple of hours on night two) and were returning from our trip to uniform issue. Having been issued our standard four sets of BDUs and dress blues, each of our flight's trainees was now marching back to barracks with a standard-issue duffle bag full of pants, shirts, socks, and Mil-spec tighty-whitey underwear. When we came to the final intersection between ourselves and our dorm, I diligently ran out to perform my duties, duffle bag whacking me in the ass with each step. I assumed my "stop" position with my right hand outstretched in front of a plain gray Camry, its driver clearly annoyed by the delay. Then, as the flight of men began crossing the road behind me, Tech. Sgt. Williams walked slowly, ominously toward me.

He inched toward me and then leaned in so close that his "smokey bear" style instructor hat actually touched my forehead (no exaggeration). He cleared his throat and then asked the one inevitable question that I had dreaded. I'm glad he did though, because I wouldn't be writing this book if he hadn't.

"Your brother was an element leader, wu'nt he?" I could feel his hot breath and small flecks of spit on my cheek. I kept my eyes front and knew that my fate was now sealed.

"Yes, Sir!" I replied.

"Be pretty cool if you were too, wouldn't it?" he asked. *Crap!* No part of me wanted to answer that question. But I did. And I lied.

"Yes, Sir!" I replied.

"Good. I'll be watchin' you!" With that, he walked away and left me to rejoin the flight, now safely across the street.

That evening we received the results of the drug test the military gives you on night one. Some people, for whatever reason, enlist in the military and then realize they're not ready for that level of discipline. I don't know if the practice has changed since I joined, but if you were dismissed within the first few days of enlistment for a medical or even drug-related issue, you were given an "administrative" release instead of a dishonorable discharge. Some of the guys knew that and had smoked pot on their "last" night as civilians. When the test came back, they were immediately dismissed. There wasn't any fanfare associated with it, you were just out.

On night one, Sgt. Williams had subjectively selected the five biggest, baddest-looking dudes to be "in charge" of our flight. The biggest was assigned as "dorm chief" and the remaining four were the "element leaders" who had been charged with keeping 15–20 other guys in line. The one of those four guys who had been put in charge of my element had "popped" on his drug test and was given the boot. Sgt. Williams made a bit of a spectacle about the changing of the guard that evening. Not because of the drug test, but because he wanted everyone to know that I was now in charge (he actually told me that later, so that's not an assumption).

But I was at a disadvantage from the word go. If you recall, I'm not a tall guy. I'm a weightlifter now, and certainly have some heft to myself, but at that time in history I was a skinny-fat, five-foot, six-inch tall "loser" who had no justification for claiming a stake to any designation as a leader. I had to fight for it. I could fill an entire other book with the journey my basic training experience took me on, but for the sake of staying on topic, I'm going to press fast forward again.

If we ever meet in person, I'd be happy to recount any of the stories from those days to you. But, let me just summarize it like this: I had to fight to prove myself as a leader. That meant I became the hardest of all five of the leaders Sgt. Williams had selected. Robledo nicknamed me "Little Hitler" because of how hard I was on my element. I took shit from no one, and they know that. In the military training environment, that style works. I caught onto that quickly and my people fell in line to avoid my wrath. I actually still have friends from those days who were shocked to realize that I'm not a royal asshole in real life, but I was playing a part to get through.

Again, those stories could fill volumes of their own. My point in even referencing those times in my life is to point out another pivotal lesson that I've brought with me along the way into the world of safety management. See, about the third week of basic training, I became terribly ill. I contracted a fairly common strain of the adenovirus and my days were filled with constant ups and downs. One day I'd have a raging fever and the next I'd be fine. The remainder of my time in basic followed that trend. I didn't realize it at the time (and if my doctors did, they didn't let on) but my condition was potentially deadly serious and has actually contributed to the deaths of trainees over the years (Venuto et al., 940).

My worst fear during that period was being "washed back" to the beginning of basic training and assigned to a different class. I was one of the alphas of my flight and simply could not stomach the idea of being knocked back down to the level of "average trainee." So I did things normal people consider stupid: I did PT (physical training) with a 103 fever. I marched with a full rucksack while I was dangerously dehydrated and hallucinating about getting set up on a date with Carmen Electra (don't judge my fantasies!). FYI, I realize I'm talking Air Force, so all you marines that are scoffing at the triviality of my suffering can go eat a box of crayons ... I know I'm not "hard."

In our fifth week, we were sent to the Air Force's "warrior week." The training schedule has changed since then, but when I went through, that was your proving ground. If you "survived" the week you were given your airman's coin and basic training graduation was essentially a foregone conclusion. There was one catch: weapons qualification.

My wife, arguably an exponentially harder person than I am in hundreds of ways, was in the Army and has recounted the numerous times she had to carry, dismantle, clean, and use her "weapon" during Army Basic Training. Air Force weapons training pales in comparison. I'm not going to beat around the bush about that. In any case, though, Air Force trainees needed to "qualify" on the M16A4 just like all other branches. That was the culmination of our warrior week.

I was a ragged, beaten down silhouette of myself by the time qualification came. That morning, I had one of my fever spikes, but I had kept it to myself. My assigned drill instructors Sgt. Williams and Sgt. Robledo, were not present for weapons training (a safety measure put in place due to the slight but real chance that someone had become so disgruntled during the previous weeks of torture that they would try to take it out on their instructors with a loaded weapon). For that reason, I had been able to feign readiness, but I was low on everything: blood sugar, motivation, energy, focus … you name it.

We practiced dismantling, cleaning, and reassembling our rifles before breaking for what should have been our lunch. Our instructors were in a hurry that day for some reason though. So, rather than breaking for chow, we immediately headed to the range instead. The one backstory element that I've not touched on here is that I was an excellent rifleman. Even though shooting is not necessarily an Air Force core competency, I had grown up with a rife in my hands. It was my solace as a teenager, having spent hours every summer day teaching myself how to breathe, control my heartbeat, and then press the trigger in perfect synchronicity with my biological rhythm. I was hungry and shaky, but I knew how to shoot and my goal was to earn "expert marksman" status.

At the range you were given the exact number of magazines, loaded with the precise number of rounds needed to qualify at that elite status. I won't try to pretend I remember what the exact score needed to be, but I remember shaking off my hypoglycemia, sighting in my rifle, and making my shots. They were true, and I could see through my sights that I had made the marks … with one glaring problem. Three of my shots had penetrated one of my five targets in precisely the same spot. It was something I had done hundreds of time in my practices at home. In fact, at one time I prided myself on being able to shoot ten rounds through the same hole. It's a difficult, but entirely attainable skill. Also somewhat uncommon. Then, the trainee next to me shot my target … twice. I actually watched it happen.

I tell that story not to brag about my marksmanship abilities, but to point out two incredibly important points: First, I was not awarded the marksman ribbon because those two rouge shots were attributed to me missing the target. It was ironic, but in the minds of the instructors, those two strays were my two "missing" shots that I had actually directed through the same hole as one I had made previously. Second, and more importantly, the trainee beside me who had shot my target did not even qualify on the M16. He hadn't hit his target the required minimum amount of times.

So how does that equate to our world in safety management? I'm willing to bet most of you have already connected those dots, but let me lay it out. Shooting at my target (not his), disqualified that airman, right? Had I not actually seen his stray shots hit my paper, no one would ever have known why he failed, only that he had not achieved his goal. How is that any different to how we determine the failure or success of safety performance in our organizations based upon an incident rate? All that number can do is point out that we didn't hit the mark; it provides no further insight (and even that is a flawed indicator based on the ridiculousness of the formula itself).

I've already laid this out in detail, but our measurement device is inherently flawed. If you still believe the absence of "recordable" injuries (or any injury for that matter) is the litmus test for determining the success or failure of your program, you're in for a rude awakening. You are shooting at the wrong target. It's time to draw the line in the sand and determine what safe means. Ready?

Stupid Simple Toolkit Item 2: Choose Your Target

If you look up the word "safe" on dictionary.com you'll find this definition: "involving little or no risk of mishap, error, etc." That's actually the third definition option, but it makes the most sense when discussing this tool. The part we need to key in are the words "little" or "no." There are far too many zealots in this field who still cling to the idea that we can eliminate all risk. Don't get me wrong, here, I'm not at all saying we shouldn't work tirelessly to eliminate every risk we can. I'm just trying to bring a little bit of pragmatism to the topic.

In the most basic sense, we have to accept that life itself is a risk. You can go on living in a fantasy world and preach about how preventable the past was, but no matter how loudly you do, nothing about it will ever change. We have to come to grips with the reality that we can't prevent all of life's bumps and bruises (very often referred to as OSHA Recordables). Until we can get past that troublesome fact we'll be distracted from what should be our real goal: not killing, maiming, or otherwise permanently altering the lives of our workers. Start by aiming at those three targets. Once you're done with that, we can talk about the double dose of Advil your company doctor prescribed for a bruise.

8

QUIT SAYING "SHALL," YOU'RE NOT SHAKESPEARE

As I mentioned earlier while retelling the Francis saga, Stacy the technical writer taught me some very valuable lessons. One, in particular, has stuck with me over the years. And while it will seem on the surface to be just an opinion or at best a good "tip," it actually has some very real application in the world of safety management. Maybe even profoundly so.

Her message was simply the statement that procedures should not have the words "shall," "should," "must," or "will" in them. Not only because the first term is insanely antiquated and out of touch with today's society, but (more importantly) they are statements of things one is going to do in the future, not actions that currently take place in the work environment. Another knock is that they are traditionally legal terms.

Though many tout their legal knowledge, safety professionals (typically) are *not* lawyers. Nor should they be. Safety professionals should be resources, teachers, and (dare I speak blasphemy) thought leaders. Too often we pigeonhole ourselves into the roles of dictators and legislators whose existence in the workplace only perpetuates malicious compliance due to the fear of a write-up, or worse being "walked off site." It also serves to further divide managers (*they*) from the workers (*us*). As I explained during the Inspector Bob saga, my struggle with that version of compliance-based "safety" has always been to question how any of it actually *makes* people safer. The answer, of course, is it doesn't.

The Stacy method was simple. Remove the legal words, provide clear instruction. Let me use the example of a screwdriver. You *could* write "Employees shall use Phillips screwdrivers to secure screws and will ensure that screw heads are not stripped when tightening." Most people will get it. Most people will do it. But here's an organizational philosophy question to consider: What happens when the employee uses a screwdriver that doesn't fit the screw (even if it's a Phillips)? Or even worse, what happens when that employee strips that screw? Now we have a compliance issue.

There are two things going on here. First, we've not provided clear direction. Employees have been told the rule, but they have no idea how to get it done. What size screwdriver is needed? How do you use the tool properly to prevent stripping? What do you do if you feel the screwdriver slip? Obviously, this example is almost too simple and probably wouldn't be as huge a disaster as I'm making it out to be. But the fact remains that we've degraded the significance of the standard by not providing a foundation on which it can be achieved. If we're being honest, how many of our safety procedures and policies are similar in tone? They can't all carry the same weight so, inevitably, "shall" becomes "eh, maybe." As time goes on, more and more screws will be stripped and more and more nothing will happen. It's a recipe for complacency.

Second, we've created a culture where workers are not empowered to interpret, plan, and make solid risk-based decisions that allow them to adjust and take action when the unexpected happens. "Shall" implies that there is no gray area or middle ground. But we know that no job site is black and white. Let's take the example one step further. Suppose a worker comes across a screw that is already stripped (perhaps because someone else put it there and didn't report the "violation"). That worker may need to use a drill and a screw extractor bit. Her manager, no doubt, is putting on the pressure to remove that old screw, and will probably even supply the drill. But a drill is not a Phillips screwdriver. Now we're between a rock and the place where the work actually gets done, even if it is done by violating the rule. My point is this: If by design the rule requires that personnel violate it to get the job done, then our objective is to catch people doing wrong, not facilitate safe, successful work. Safety should be the resource that others lean on to solve the complex problems they can't figure out. Our directions should provide the defenses that enable them to get work done without getting hurt. We don't have to complicate it. How much more effective would our instructions be if we stated requirements like this:

- Select a screwdriver that is sized to fit the screw you will be installing/uninstalling.
- Turn the screw clockwise to install, counterclockwise to uninstall.
- Prevent stripping the screw head by fully seating the driver into the screw head and applying only enough pressure to turn it.
- If the screwdriver slips, stop and readjust. Do not force it.

Some who read this may think I'm being nitpicky or pedantic, but I think this is an incredibly serious point. If, through the language we use to "manage" safety we alienate, confuse, remove accountability, and drive non-compliance, I would argue that we're either really terrible at our jobs or just plain terrible people (I would also argue that some are both). For the love of God, stop trying to impress people and give them tools they can use with clear expectations of how they're to be used. In short, say what you mean.

My experience with clear communication in my own career has some direct parallels to this concept. During the twilight of my time in purgatory alongside Francis and THE FOUR, my team was assigned a "new" supervisor. Dave had apparently ruffled feathers with our new manager and was relegated to a small office in the corner of a trailer on the edge of the plant site. In his stead, we were blessed with another who I will refer to as "The Tongue."

The Tongue was an incredibly skilled industrial hygienist by trade, but he was grossly unqualified to interact with humans on any type of personal level. To give you an example, he once physically pulled my (manager approved, personally owned) safety glasses off my face because they were Oakley brand and he did not believe that Oakley made any ANSI compliant, stamped safety eyewear. He also thrived in pitting members of the team against one another; but those are different stories for a different book.

Our site (company actually) used an online database for tracking all safety statistics. It was used for everything from training, audit findings, safety initiatives, and (of course) incidents. The system had some quirks, but it was pretty robust and very valuable from a reporting and statistical analysis standpoint. One key requirement for the administrators of the program at each project site was that all incidents (regardless of severity) be entered into it monthly.

Most sites only had one administrator and there were no overarching rules or standards for how or, more importantly, *when* incidents were entered into the system as long as they were in there by the fifth business day of each month (for the prior month). At the chemical plant there were three administrators, myself included. If you recall, the plant was in startup mode at this time and we were still figuring out how everything worked. To his credit (and believe me, he deserved very little), our new supervisor noticed that each of us had our own particular style. Joe, for instance, would enter events as they happened. Tommy would wait until each investigation was complete. I, being the black sheep, would let the incidents pile up on my desk and enter them in one fell swoop on the first business day of the new month. Seeing these inconsistencies as a potential opportunity for mistakes, he pulled the three of us together for some "coaching."

What that actually meant was that he simply told us we needed to "get our shit together" and start doing the entries right. I immediately fired back and explained that we all had different styles based on expectations from previous projects. I courteously asked him to provide us with the expectation for our current site, and to my surprise he appeared receptive to that request. The others helped by backing me up and reiterating that we simply wanted to know what our current management team wanted. The Tongue committed to asking our manager Wally (coincidentally his best drinking buddy) and getting back to us in the morning.

At 6 a.m. the following morning, I walked into a stoic and uncharacteristically quiet group. The Tongue glared at me as I made my way in and took my place in the circle.

"You got my ass chewed last night, SMASHMOUTH!" he pointed a finger at me before I had even settled into my chair. I don't remember what (if anything) I said in response, but I'm sure my mouth hung open as he continued. "What kind of dumbass doesn't know how to enter data into a computer? Huh, SMASHMOUTH?" I was baffled. I had no idea why he was calling me that or what it meant, but all I could do was loop the lyrics "hey now, you're an all-star" through my mind on nonstop replay. The meeting continued for what seemed like hours and all of it was centered upon my "SMASHMOUTH" and how stupid I was for asking a clarifying question about expectations. It was a downright shame and actually the straw that broke my camel's back. I quit that job on the spot that day, just one week shy of being promoted from a junior to a senior safety engineer. A raise that would have taken me into six-figure salary territory.

I tell that story for a couple reasons. First, because I still find it incredibly bizarre, and even mildly funny. But second, because I think we safety "pros" communicate to our customers in confusing hyperbole far more often than we realize and certainly more than we should. People can and will do extraordinary things, but everyone needs guidance at times. That guidance should always be clear, concise, and simple. Consider this next time you're explaining a safety issue or requirement to someone: if you gave the instruction once and then walked away, could you be certain the job would be completed to your specifications once you were gone? Most likely, the answer is no. We have a tendency to explain things from our perspective rather than meeting the other party at their level of understanding and knowledge.

With that knowledge in mind, do this one (stupid) simple thing *every time* you ask someone to complete a task. Ask them to show you *how* they will do it.

Stupid Simple Toolkit Item 3: Write Better

Standards, Procedures, and Guidelines – A writer's reference:

This will probably be something I end up taking on as an entirely new project at some point, but since I'm advocating for it I owe it to anyone reading to at least give you some thoughts here. It's not brain science, so don't expect your head to explode.

- **STANDARDS** – These are the minimum requirements you expect anyone at your facility to live by. **DO NOT** (I cannot stress this enough) simply regurgitate government regulations and call them "your standards." Aside from being incredibly lazy, there just isn't any value in this practice. I understand that this is what most people do, but think of it this way. If everyone needed to have the ability to interpret OSHA, what need would they have for a "safety professional"? Here are your rules to live by:

- o Avoid exposition about all of the "bad" things that could happen if the requirements are not followed. That's what training is for.
- o Provide clear, direct statements about what is required. For example, you could write "Employees are required to maintain 100% fall protection when working at heights above four feet."
- o Never prescribe specific methods for executing a requirement (that's what procedures are for).

- **PROCEDURES** – These documents provide your employees with clear direction about how to comply with a standard. Here are your golden rules:

  - o Never use the words "shall," "should," "must," or "will." (I'm a dog with a bone, did you think I'd give that up?) Just tell people how to perform the task. Use directions such as "do this" or "do that."
  - o Provide descriptions of upset conditions that will alert the user of the need to stop and adjust.

    - ▪ These are items that should be trained diligently.

  - o Partner with the people who will use your procedures and ask them how they could best comply with the requirements (standards). Your chances of creating something useful increase exponentially when you do this rather than try to legislate from behind your desk.

- **GUIDELINES** – These documents are tools that provide helpful tips and best practices. Use them to add value and impact, not to circumvent the more difficult process of writing a standard or procedure. Do it this way:

  - Provide examples of what works and what doesn't. Use pictures and step by step analysis of the lessons learned that lead to the creation of a particular guideline.
  - Reference any relevant **Standards** and **Procedures**.

There isn't a lot to it, but that's the point. We spend so much time trying to make our work seem more intelligent and regal than it needs to be. Make it simple and to the point and you'll get more mileage out of it than you could imagine.

9

ADD VALUE, NOT WORDS

Jerry, as anyone could clearly tell, was not a stable man. He was a loud, Copenhagen dipping, whiskey drinking hot head and he was one of the first people to "train" me once I got out of the Air Force. He was an "old-timer" who had been with the company nearly thirty years. There were actual legends of his exploits on various projects, and there was no denying he had an outstanding résumé. He was going to transition to a new project soon but had been placed in charge of our small team until our newly assigned manager (Nick) arrived on site.

Two weeks into my new career, Jerry gave me a very serious assignment. I was to gather up the visitor safety gear (hard hats, boots, glasses, etc.) and be at the safety table at 9 a.m. Then I was to meet some important person named Joe, or Bill, or something. All I got out of the rushed order was that he was a big wig of some sort. I was to give him a safety briefing before he would be allowed on site. More specifically, "Bill can't go one foot on that site until you give him the safety briefing and he signs off on the visitor orientation sheet!"

With that Jerry left to go do his rounds on site and I gathered up my supplies. I was dutifully positioned at the table at 9 a.m. when no one showed up; 9:15 rolled around, then 9:30. No Bill. Jerry walked back in at 9:45 and came unglued.

"Bill didn't show?" He seethed.

"I haven't seen anyone since you left," I replied. Having only been an employee for two weeks, and not having any clue who Bill was, I had decided staying put was my best chance at getting my briefing done when he finally showed.

"I know where they are. Come with me. And grab those boots and hats." With that, Jerry began walking toward the elevator. I followed, haphazardly carrying our visitor's safety gear.

Once on the ground floor of the office complex, Jerry cracked every door and peeked inside. Most of that floor was a maze of conference rooms and small meeting spaces. Finally we came to room 1A. It was the "grand ballroom" of all our gathering areas, complete with a makeshift stage and auditorium seating. There were two ways into this all-important conference room: the back door, where one could inconspicuously tiptoe into an in-progress presentation, and the front door, which lead you directly to the stage. We could hear that there was an assembly in the room, so Jerry didn't even bother cracking the door to check. He burst through the front door, just as the project manager had clicked forward to the next slide of his meticulously crafted "project overview" presentation. Theater lights blinded both of us and time stopped in the room as Jerry cleared his throat and motioned for me to follow him in, stack of boots obscuring what little vision I had left.

"This is Jason," Jerry bellowed. "He has a safety briefing to give you and you're not allowed on site until he gives it." Jaws gaped across the room as the project manager composed himself quickly and motioned for me to take a seat in the back. I felt like the kid who has recurring nightmares of standing in class naked as I shuffled noisily and then clunked my gear down on the table. When I looked up Jerry was gone and the project manager was reorganizing his thoughts before continuing on.

Then, I sat in embarrassing agony as I listened to the remainder of the hours-long presentation, the whole time dreading the fact that I would actually have to give my safety briefing once it was over. The punchline of the story is that Bill, then COO (future CEO), was actually a very pleasant man. He remembered me on a first-name basis for the entire time I worked for that company. I wouldn't be surprised if he still did.

I don't believe Jerry even knew (or cared) what had gone down that morning. He didn't care that he had interrupted a high-profile presentation which had been meticulously planned for months. He certainly didn't care that he had put me on the spot and made me temporarily doubt my new career choice. He didn't even care that the man he had called out in front of a crowd of seasoned professionals was the second most powerful person in one of the largest, most successful companies in the world (yes, world ... you read that correctly). He only cared about the one thing that concerned him, getting a useless, Cover Your Ass of a "safety briefing" signed off. He was one of the most narrow-minded individuals I've ever worked with.

Here's the thing. There's nothing inherently wrong with that type of laser focus. As with many aspects of success in life, it certainly has its purpose. Sometimes "laser focus" is the difference between success and failure. But when that focus betrays your objective, you have a serious problem. I don't believe Jerry understood that. If he were the only safety person who suffered from that problem, our profession would be just fine. But he's not the only one.

So many of us get so wrapped up in what is directly in front of us that we miss everything happening outside of that self-imposed tunnel vision. I call this being "singularly focused." Those with true leadership ability are able to focus on what's in front of them, while also realizing that everything else happening to their right and to their left plays a role as well. Before I get too philosophical about all that let me illustrate with another fun Jerry episode.

I've already mentioned that the man was a little ... er ... a lot unstable. What I mean by that was he was outrageously emotional. If "safety" was involved in an issue and Jerry had anything to say about it, that issue was _deadly serious_ (even if it wasn't). The problem with that mentality is that when everything is an emergency, nothing is. But good luck trying to coach him on that concept. He had 30 years of experience, by God! Just try telling him he was wrong.

He had a terrible habit of flying off the handle at a drop of the hat any time he witnessed something he deemed unsafe. His diatribes frequently escalated into all-out, red-faced screaming sessions where he would berate a worker for their poor choice and willingness to violate the rules just to get the job done. I'm not exaggerating the point, that's just how he conducted business. Anyone who's worked in the safety field has likely met more than their fair share of Jerrys.

His antics came to a head one day in particular when he noticed a Hispanic worker on our site who was wearing an extremely worn and faded hard hat. Those who have worked in the construction field know that a man's hard hat is often the trophy he carries with him from job to job. Often workers are given a sticker that signifies the training or time spent in each location. Everyone knows that hard hats have a limited life span in terms of their ability to protect a worker's head, but that doesn't change the fact that many attach each of those stickers from each job and wear that helmet as a badge of honor and proof of hard-earned experience.

This man was one of those. His hat reflected decades of experience and he was exceedingly proud of it. Jerry couldn't have cared less though, because he noticed the hat was cracked and brittle. He knew that it wouldn't serve its purpose if put to the test. As soon as he had made that determination, the confrontation was on. He (physically) grabbed the man's hard hat from his head, rambling about how it was worn and ineffective. He berated the man in front of his co-workers, and then directed his anger toward the foreman who had let this atrocity take place.

Everyone watching the spectacle was in shock. I knew instinctively that the worker spoke little English and understood almost none of what Jerry was trying to say. His reaction was not one of anger, but of fear (maybe because he thought he would be fired) ... and then one of sadness. Jerry, realizing that his "message" was not being received then performed the standard "squeeze" test on the man's hat to show that it was in poor condition. If you've ever performed this test on a good hat, you know that it really doesn't amount to much more than watching a piece of plastic flex and then return to its original shape. As you can imagine, however, this hat was different.

It *snapped* into two pieces directly down the center as soon as pressure was applied. With that separation I literally watched a man's legacy crumble before his eyes and his soul become crushed. Every sticker, from every job that man had ever worked for more than half his life, was on that brown piece of plastic. In less than half a moment that legacy had been destroyed. Aside from the horrific injuries and fatalities I've unfortunately been exposed to, very few moments in my professional life have been more gut-wrenching. None (at least that I can recall) have been as unnecessary.

Jerry had a point to make that day, that's not worth arguing. No one will ever be able to say that worn hat *would* have made that worker unsafe, but I can say with certainty that the lesson that could have been learned at that moment never was.

I had a short exchange with our project manager regarding the incident not long after in which I had conveyed my distaste about the whole thing. His response to my concerns was something along the lines of "Well, but ... Jerry has a lot of experience to offer."

I remember looking him square in the face and telling him point blank that he was wrong. "He's got nothing to offer," I said. "Because as soon as he opens his mouth, everyone quits listening."

How many times in your career have you had the opportunity to beat someone down versus the choice to build them up? You can use any words, or any communication strategy you want. But how much of what you say makes an impact? I think that's something worth seriously considering.

Stupid Simple Toolkit Item 4: Be Useful

Have you ever been so passionate about something that you want to talk to everyone and tell them how interesting it is? If not, that's OK. Not everyone has that type of personality. If you're struggling to imagine what I'm describing, just think of anyone you've ever met who's into CrossFit.

My point is that everyone takes pride in something. Within industry, I've found that most people who do a job for any length of time are proud (to one degree or another) of the work they do. There's a simple way to bring this out of anyone that I actually developed as a defense mechanism for my lack of experience when I first entered the civilian world. I quickly learned, however, that it was something profoundly useful. And easy to boot.

Safety professionals (and managers in general) have a habit of walking into a work zone with the goal of finding fault. Deny that fact if you want, but it's human nature to try and fix what's wrong with others. It doesn't matter if we do it because it makes us feel better, or useful, or any other reason. We do it. Maybe we should blame "Sesame Street" for constantly asking us which one of these things is not like the other. I have no idea.

So here's your stupid simple tool. It will put you on level ground with the person you're "watching." The next time you're walking around and notice something weird, or out of place, or maybe even downright egregious, approach the person doing that task with genuine interest and ask this question: "What are you working on?"

The worker (let's say he/she is a 25-year veteran welder) will likely say something akin to, "I'm welding stuff." That is to be expected. It's also the point where you flip the script and level the playing field. It's your opening.

Now, swallow your pride, pretend you're a humble person and say this: "Nice. That takes some serious skill. I wouldn't even know how to start that job. Could you show me how you do it?" Obviously, you can paraphrase and adjust the wording a little, but the key is to get them talking about what they do. The ones who are proud of that work will let their defenses down (if only for a moment) and tell you where that pride comes from (hint: it's because they have a skill you don't). Once you break that wall, the possibilities are endless.

Here's the kicker: Maybe you noticed something wrong with the way they were welding. Maybe they were doing something unsafe. If you had started there and told them they were violating a rule, that worker would never trust your motives. They might change their "at-risk" behavior for the moment you're watching, but as soon as you're gone, their motivation to comply will be as well.

If you take 30 seconds to find some common ground, you'll be amazed to see the kind of change that can take place. The only caveat to this principle are those rare, one in a million, times when someone's about to chop off a limb or fall off a ledge. If that is the case, scream at the top of your lungs and do whatever it takes to make sure that doesn't happen.

10

CHOOSE YOUR ENEMIES WISELY

When Nick showed up, he didn't make any immediate judgments about who or what each member of his new team was, he just took us at surface value. He didn't try to change our dynamic right off the bat either. It's actually a management strategy that I've tried to adapt myself. I call it the "look, listen, and feel" (you know ... CPR training) method. Essentially all that means is you take the time to scope out your surroundings whenever you're in a new environment rather than coming in guns blazing and try to change the world. It usually works pretty well. Usually.

In our case, however, Nick missed one key problem. Kelly, as I mentioned from the start of this thing, was a consummate people pleaser. Between the three of us (Tony, Kelly, and me) he was the oldest by nearly twenty years. I always suspected that rather than that fact being a strength, he considered it a chip on his shoulder because he was not at the point he wanted to be in his career. As a result, he frequently did things that served himself rather than the team. To be fair, Nick caught onto it fairly quickly, just not quickly enough.

Without knowing either of us from Adam, Nick made a decision that haunted me the entire time I worked for him. We eventually laughed about it together years later, but it never came out while he was my boss. The situation was as basic as they come.

At the time I was still shaking off my Air Force mentality and trying to figure out how my new civilian career worked. I had just witnessed Jerry literally "crack skulls" and was trying to figure out what being a "safety guy" even meant. In the military, you did safety because it was required. There were no questions, no debates, no gray areas. It was an order. Smart people don't question orders. All I had seen thus far in the "real world" indicated that things were much more ... squishy. Yet they were no less expected. I had no idea how to manage that conundrum.

On this particular day, I had been venting to Kelly about an interaction with a particularly difficult contractor. The contractor and I had been watching a boom-lift operator navigate a steep grade with his 65-foot boom fully extended. Aside from the grade, the soil beneath the lift was loose and not well compacted. I had requested that the contractor instruct his operator to retract the boom before traveling (actually a regulatory requirement, but what difference does that make, right?). His answer had been a simple, yet firm "No!"

In that moment, on site, I bit my tongue. I let it go even though both of us knew that meant I was accepting defeat and allowing something wrong and dangerous to continue. It's a feeling that I've forced myself not to forget and worked tirelessly since to never feel again. At that point, though, I had not learned how to defeat obstinate people with their own arrogance (as I would do to Francis later). When I had returned to the office, however, I unloaded.

In a gesture I considered at the time to be empathy, Kelly lent a dedicated ear to my frustrations. I had a bad habit (actually I still do) of retelling stories by inserting the words I thought or wished I had said at a particular moment in time without making it clear that I had actually maintained my composure. I'm actually very good at self-censorship when I need to be, but I often fail to explain the distinctions between what I *actually* said and what I *wanted* to say when I'm worked up about something. This was one of those times, probably one of my worst to be honest. Kelly took my retelling of the episode as gospel and (unbeknownst to me at the time) was "very concerned" about the way I interacted with people in public. He dutifully reported those concerns to our new manager as soon as he had "talked me off the ledge" and helped me cool off.

Nick, sensing Kelly's strong desire to grow as a leader, assigned him the task of counseling me to improve my interpersonal communication skills. Kelly was unaware that I actually have a bachelor's degree in that, but whatever, I'm not bitter. He sprung the session on me as I walked past the downstairs lunchroom a few days later, ushering me to sit down next to the softly humming soda machine. In his hands he held an orange book with a picture of a globe and some fancy title about how to win in your career. I'm not going to actually call it out by name because the truth is I never read it, mainly out of spite. I have no idea if it really was BS or not, but thinking about it still makes me shake my head.

I won't try to recount that counseling session either, because I have no idea what was said. I was boiling, not listening. The long and short of it, though, was that Kelly was convinced of one thing: I took things too personally and needed to learn how to let stuff go. That meant Nick was convinced too. In fact, he cited that as my number one needed area of improvement for the next three years in each of my annual reviews. Kelly had actually quit the company and moved on before I finally pulled that stupid orange book out of my bottom drawer and explained the real story to Nick years later. That moment itself was a highly important lesson on timeliness and making sure you're understood for what you really are, but that's not what we're discussing here, so I'll just leave it at that.

The real Kelly lesson came months later. He was about to embark on a vacation that he and his wife had been planning for the past year. In his stead, I was asked to fill in on his contracts. Since I haven't explained this previously, it's worth noting that we were part of a contract oversight team whose responsibility was to enforce compliance with OSHA law (obviously), but also with contract requirements. We did not actually employ any of the labor force that was performing the work. Our impact on worker safety hinged entirely on our ability to ensure that the contractors weren't playing cowboy and taking unnecessary risks. The only teeth we had in that regard was by pinching their wallets in order to get them to fix hazards.

By this point, Nick had split the various individual "contracts" between his team of three, although not evenly. I was still very green and only had one (the smallest one) of the many jobs assigned to me. That meant that most of my time was spent doing administrative tasks. Not doing the "real" safety work out on site. I wanted to be out in the field like the other two more than anything, so you can imagine how elated I was to get this opportunity to cover. The day before he left, Kelly took me out on site to show me the ropes.

His main contract was a multi-level garage which was progressing at lightning speed. The contractor was well ahead of schedule and aiming toward a huge bonus (I'm talking millions of dollars) if they completed construction early. The structure was planned to include six levels of parking. At their current pace, they were adding a new level every two weeks by completing monolithic concrete pours once a week, each pour making up half of the new level. The hitch in this process was that each new level needed to be supported structurally by what is referred to as falsework. Essentially it was a web of steel, wood, wire, nails, and other various braces meant to hold the new concrete floor in place until it cured. Once that occurred, the falsework could be removed, but it needed to be in place until that point. If you're not familiar with the craft, envision trying to pour a bucket of epoxy (glue) into a mold in order to make it into a flat tabletop surface.

Per contract, the falsework for each new level needed an expert to "sign off" on its structural stability before concrete could be poured. Kelly proceeded to show me how all of it worked and to describe what it was intended to do as we walked through the construction site. It was all extremely interesting, but even now I'm not certain I can explain it with the level of authority to say what would or would not be acceptable (structurally speaking). Then he showed me the form.

Kelly, in his desire to please, had been placing his name in that blank space that required an "expert." He had been *certifying* (for months at this point) that the installation of these supports would prevent incident, injury, even death. I may have been green at that point, but I knew Kelly was no structural engineer. I knew he had been potentially signing his life away. I also knew I was not willing to join that club.

For those who haven't worked in this field who may not understand the gravity of what I'm explaining here, let me lay it out clearly. This isn't about money. This is about murder! That's not even me trying to be provocative. The long and short of it is simple: If that falsework had failed and killed someone (a distinct possibility), Kelly would have been held criminally responsible because he had guaranteed that would not happen just by writing his name on a piece of paper. Forget the impact it would have had on our company.

I told Nick about it immediately and he knew then that we had a weak link in our team. Anyone who failed to see the monumental nature of what had been occurring over those months was no one who could be trusted. He shored that problem up better than any manager I had ever seen previously and have ever seen since. From that moment on our team changed with one swift, distinct move.

Stupid Simple Toolkit Item 5: Build Your Safety Table

Nick called what had been happening "playing Mom against Dad." Kids do it to unsuspecting parents all the time in order to get what they want. Usually, it amounts to that kid getting an extra serving of dessert or getting to stay out later with some rebel friends. Aside from the moral implications of taking advantage of someone, there usually isn't any lasting damage. When leaders do it in regard to industrial safety, it amounts to a life and death stakes bet being placed against the lives of their workers. They see it as getting past the "roadblock" of safety to the job done. In all honesty, it usually works in their favor. Until it doesn't.

This is another of those areas where our actions as safety practitioners can either define our real purpose or perpetuate the stereotype of being a frustratingly unnecessary extra step. Ours is not a charge to throw up barriers and hold the line until every "t" is crossed and "i" dotted, it is a responsibility to partner and help people safely through their obstacles. That may sound simple, but those who understand know it is a truly delicate dance where there are certainly lines that cannot be crossed. It's not always pleasant, which is why it is tempting to just sign the paper and turn the other way.

No matter if you're part of a team, leader of one, or a solitary operator your effectiveness as a safety resource can be threatened by anyone's desire to "just get it done" at any time. You might even be amazed by the strong characters who are willing to bend just a little too much when they're up against a wall. In order to keep that from happening you need information. You need to know what's happening and what the stakes are. It takes skill, but you need to have your hand on the pulse of your business. That means being involved in it.

Nick accomplished this by one simple decree: The team will meet every day at the safety table at 11 a.m., no exceptions. For us, the method worked. We each shared our knowledge of the parts and pieces we had access to and then made decisions as one. No one was allowed to individually deviate from any decision made at that table. At first it was clunky, but once people realized there were no independent operators on that team, there were no more opportunities to pull a fast one.

So here's your challenge: Build your own safety table. In your organization that may mean something entirely different than Nick's design. But the point is this, establish yourself (or your team, if applicable), as an unwaveringly solid object. When others understand that your integrity is not flexible, they'll stop trying to find its weaknesses. With that nuisance behind you, you'll be able to get to the real work required to help your organization attain the solutions they so desperately need.

11

THE ONE WIRE DIFFERENCE

Nick called the safety hotline that September morning at 7:30. I usually carried the team's state of the art Blackberry Curve during the weekdays in case a "safety flash" needed to be emailed out. Mainly because Nick couldn't figure out how to use the thing and the other two were in and out of service depending on what part of the construction site they were on. But it was odd for Nick to be the one calling. Ordinarily, if he needed me for something, he'd text me on my personal phone. This morning was different.

"I'm sending you some pictures. I need you to send out a one-hour to corporate. We had a really bad incident this morning." My heart skipped a beat as he said that. Needing to send a one-hour meant someone, maybe even more than one someone, was being transported to the hospital with serious injuries. Maybe even worse.

"OK," I replied. "I'm finishing up my appointment now. I can do it from the car." I had been at the doctor that morning which was the only reason I wasn't already on site seeing what had happened in real life. I waited in silence for what seemed like an eternity. Nick hadn't told me what had happened, so I was left to my imaginative devices. My head usually goes to the darkest places first, so I was not looking forward to the pictures that were inbound to my phone.

Finally, the flash of a new message popped up on my screen. I opened it and waited for a few more agonizing seconds for the image to load (for those who don't remember, Blackberries didn't perform like what we have today). The picture was grim. It showed one … two … five men literally crushed inside the collapsed structure of a rebar caisson (for those unfamiliar, a rebar caisson is the metal structure that forms the structural support on the inside of concrete pillars used in road and bridge construction). First responders were using all manner of tools and emergency equipment to extract the men from the folded pile of steel. Beyond that, I couldn't even determine if they were all still alive. I called Nick back to get the rundown and make sure I didn't write anything I shouldn't in the one-hour notification. Time was ticking, but this was one of those moments where you needed to be certain of your word choices. Careful wording in these types of reports is required just as much for legal purposes as it is for managing the fallout.

Nick described the incident in detail and then hung up to let me think. He had absolute trust in my ability, but he also knew I have a strange writing process. That kind of trust from a leader is rare, and I did not take it lightly. We both knew that by the time I had drafted my one paragraph report I would have only seconds left to meet our 60-minute deadline. Nick would not have time to review my work before it was sent.

I sent a short description in that first report similar to the words below. I've edited out anything proprietary or legally binding, but this is what happened:

> A 70-foot by 6-foot rebar cage was being assembled by six iron-workers who, at the time of incident, were on the inside of the cage. The cage collapsed, trapping five of the six workers. Rescue operations followed resulting in all five of the trapped workers being transported to the nearest local hospital. No additional information is known at this time.

I still have the picture (actually pictures) I've described here. Obviously, I won't say where or when this happened, but that isn't really a relevant factor in telling this story anyway. The most important part wasn't discovered until later on that day during the initial investigation. In the pictures, one can clearly see at least two completed caissons laying on the ground next to the pile of rubble which nearly claimed five lives. That meant that successful completion of the task was not only possible, but likely. The very crew that had been injured had already done it. And not only on our job site. These were skilled professionals.

It would be easy to attribute their failure to a solitary poor choice, or "at risk" behavior, but that's simply short-sighted, to put it lightly. During the investigation, the contractor supplied a detailed procedure which had been trained and practiced for years. It clearly explained that the process for completing these giant structural supports was to be done in phases:

1. The center supports, which resembled a wheel with eight spokes (see Figure 11.1), were to be lashed together and then spaced out equally along the 70-foot length of the intended steel cylinder.
2. One 70-foot beam of rebar at a time was to be lashed from the bottom up around the center supports, forming a half-moon shape. Workers were allowed to do this while standing in, around, or above the new construction.
3. Once a half moon, or cradle, was formed, workers were expected to exit the inside of the structure. The remaining 70-foot rebar was then added and lashed to the outside to form a complete cylinder. At this point, the structure was a wobbly and unsecure thing which was prone to laying down on itself (think of a slinky laying on its side).
4. The final step in the process was to wrap the entire 70-foot span of the cylinder with individual rebar coils spaced about 6 inches apart. Each bar was lashed with tie-wire in multiple places which stabilized the structure and prevented collapse.

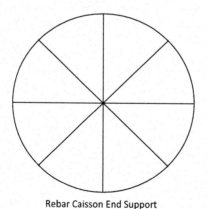

Rebar Caisson End Support

Figure 11.1 Diagram of the center support

I'm sure there are other ways to build a caisson, but that is how it was done on our site at that time. On the day of the incident, however, something changed. For some reason, the crew had deviated from the process. They had disregarded the instructions in step 3 which required them to exit the structure and had proceeded to build over their heads while remaining inside the steel cylinder. At each end of the structure one of the star-shaped center supports stood like a silent gatekeeper, requiring contortion and flexibility to bypass.

Due to that inconvenience only one man, the foreman, was required to squeeze himself through the opening (you know because time is money) in order to shuttle tools and equipment to the other four men working dutifully inside. At some point, that inconvenience became too much and he decided to remove the obstacle. He pulled his wire snips out, cut one wire on the outermost center support, and the entire 70-foot-long, 6-foot-diameter steel tube fell over on itself, trapping the crew inside.

A crane operator who had been performing a pick just behind the rebar crew's operation heard the commotion and looked behind him through the rear window of his cab. Had he not reacted quickly enough and turned his mobile crane 180 degrees around, four of the men inside that pile of steel and wire would have died, though the foreman only had his leg pinned (albeit broken in several places). The operator extended his boom over the twisted mess and was able to position his block and tackle in just the right spot to remove just enough weight from the men trapped inside. There was no escape though. Since the structure had not been stable, each bar bent and flexed as it pleased and the caisson could not simply be picked up as one piece. In the end, emergency crews had to cut individual exit holes for each man. It was an agonizingly long rescue.

I remember walking beside the collapsed caisson by myself later that afternoon. The emergency vehicles were long gone. The injured were beginning their long road to recovery (though two of them were changed permanently). Most of the site had packed up and gone home for the day. A pickup or two would happen by every now and then, but I was the only one left exploring the rubble. I don't recall questioning much as I stood there because there really wasn't much to question. What had occurred had been figured out rather quickly. I don't believe we will ever know why someone on the crew made the decisions they made or why procedures were not followed, but in the end knowing any of that could never alter what happened.

That moment was a turning point for me philosophically. Even if I couldn't quite articulate it, I was beginning to realize that our mistakes (or lack thereof) are not what determine our success. That holds true when speaking about safety or any other area of life for that matter. The real key is making sure there's enough "wire" holding your process together.

Stupid Simple Toolkit Item 6: Add More Wire

In his book *Just Culture*, Sidney Dekker states "There is almost no human action or decision that cannot be made to look flawed and less sensible in the misleading light of hindsight. It is essential that the critic should keep himself constantly aware of that fact."

I reference those words in relation to the story you've just read because it's critical that we look past any of the "choices" that were made by anyone involved that day. Unless we can do that we'll be perpetually chasing our tails like so many who have touted their snake-oil "behavior" improvement safety programs. We can either keep drinking that Kool-Aid or decide to do something else that actually makes a difference.

From this point, the tools get harder to use. Because using them will put you in a paradoxical situation. The things I'm advocating here are no longer just tips you can use to communicate better or provide better training and understanding. They're proactive measures. That means we're going to start attacking the hazards that can affect our people. Every time you do that you are guaranteed two things:

1. A hazard, which could cause harm, will be removed;
2. That same hazard is now gone, and you'll never be able to prove it would have caused harm.

So here's the tool. Look critically at your organization, your equipment, your processes, look at all of it. What I mean by "critically" is not a challenge to figure out who screwed whom or who sucks at their job. It's a challenge to find weaknesses. Find those places where only "one wire" stands between your people and catastrophe. Why those situations exist is irrelevant. What you do about them can mean the difference between life and death.

12

READ BETWEEN THE LINES

One May during college, some friends and I decided to spend a Saturday afternoon at the movies. We chose a film that had been out since March, assuming that the length of time it had spent in the theater meant it *had* to be a good movie. We could not have been more wrong. It was so bad, in fact, that the geniuses in Hollywood made four sequels. That should tell you something.

After the previews rolled, the film's opening credits began only to be quickly and violently interrupted by deafening white noise and bright flashes of light on the screen. As the noise grew louder, we looked up to the projection booth and realized that no one was there to fix the problem. A few people got up to leave as one of my friends went to alert the theater staff. The rest of us stayed in our seats and waited for the movie to resume.

Suddenly the blood-curdling sound stopped. The screen went black. We heard two clicks before the previews we had just watched restarted. And then ... white noise and bright flashes of light! This process repeated two more times before the reels were switched and the film played as intended. What followed was a horrible movie about a high school kid who has a premonition of terrible things to come and saves his whole high school class from a horrifying plane crash. Though he could have dismissed the vision as a daydream, he had the sense to act upon the warning he had been given (they all die in the end anyway, but that's beside the point). Had I been wiser at the time, I may have been able to notice that the technical difficulties preceding the film were a warning, foreshadowing the excruciatingly awful two hours to follow. How's that for cruel irony?

My point in all of this is simple: Life is full of indicators, lessons, warnings, and other signs that help us navigate our way safely through each task we take on. I don't expect too many vivid premonitions, but every time I experience an accident or injury in my personal life, I'm able to pinpoint what went wrong almost immediately (remember the third option for dealing with risk?). Even worse is that most of the time I knew there was a problem or unacceptable risk before the incident occurred, but I continued on anyway. Having conducted hundreds (maybe thousands) of incident investigations throughout my career, I've noticed that people perceive an element of knowing what's going to go wrong before it does very often. As I've already discussed ad nauseam, I believe a lot of that is our hindsight filling in blanks we couldn't see in the moments before something went wrong. But the thought process does beg a serious question: How many obstacles do we leave in our way which, if removed, would not cause harm?

In one of my shorter contract-based roles, I spent a great deal of time conducting your run-of-the-mill safety training, providing safety, bulletins, and generally encouraging the company's supervisors to do the same. I won't make it out to be more than it was, I was just there to check a contract-required box. We've all had similar roles I'm sure. For what it was, though, the job was at least tolerable because most of the supervisors were bought in.

One supervisor, though, was extremely negative about all things safety and constantly pushed back. He argued that we talked about safety too much and all of it just went in one ear and out the other (one might argue that his crew didn't listen because of his delivery, but that's a different discussion). His exact words were "No one comes to work to get hurt, so we shouldn't have to remind them every day." While I do tend to agree with him that no one *wants* to get hurt, I question how often any of us approach a new job or task and actively participate and try *not to get hurt*. Think about that the next time you grab a hammer and some nails to hang a picture on your wall at home. Do you wear safety glasses? Do you stand on the couch instead of moving it and getting a ladder? Do you even wear shoes?

I find it interesting that people can be so zealous about the rules and regulations when there's money (work) involved, yet shirk those "responsibilities" as soon as we pull out of the company parking lot. Don't get me wrong, I've been guilty of it at times too. Usually when my son is standing right there ready to call me out on my unsafe work habits and remind me what I do for a living.

On some level, whether at work or at home we're all motivated to get things done and accomplish good work. There's absolutely nothing wrong with that … unless that motivation overshadows your desire to do right by the people you support. When that happens, we miss our opportunity to pause, assess our environment, and listen to those indicators that are warning us of undesirable consequences.

Many of you probably already employ a process like the one I'm about to suggest, but if not try to get this habit started with your people. Before they start each task ask them to STOP, look around for a full 60 seconds, and use their skills and talents to identify the things that could cause harm or do damage. You'll be surprised how many of those things are under your control. Once you figure out what they are, do something about it. The process doesn't need to be written out on a form, it needs to be a habit that begins teaching people the difference between just saying "Be aware of your surroundings" and "Be an active participant."

I know what you're thinking right now, but don't write me off just yet. What I just described might sound eerily familiar to parts of some of those "snake-oil" behavior programs I just pooh-poohed in the previous chapter. That couldn't be farther from the truth. What I'm advocating here is training, planning, and empowerment. Our workers are our first, best resource for finding our "one wire" problems. We need them to participate and must expect that they identify those items. When they do, it must be explicitly clear that our expectation is that the problem be fixed before they move on.

Stupid Simple Toolkit Item 7: Teach Active Participation

I've already alluded to this, but one of my pet peeves is hearing a manager or (worse) a safety professional tell someone that they need to be "aware of their surroundings." To me it's one of the biggest cop-out Cover Your Ass methods that can be used. Figuratively it amounts to wringing your hands together and walking away before something actually happens just so you can feel good about yourself. It's the safety equivalent of using hope as a strategy.

I recently presented this concept to all of the employees at my facility in a little bit of an unconventional way. During a series of safety meetings I pulled up a random participant from the audience. At the front of the room I placed a small 8-inch-tall step stool on the floor and gave the following instruction:

> Please step up and down on this step five or six times. Be careful you don't fall.

The audience watched as nearly every person who was asked to do this task stepped up and then back down in nearly identical fashion. First, the person would place their right foot on the step, then bring up their left to meet it. To step down the person would shift their weight to their left foot and then step back down with their right. Some varied the process a bit by taking the next first step with their left foot, but the result was always the same. Once on the platform, the person shifted their weight and stepped back down with the first foot that had been placed on the step. Nothing shocking there, that's how normal people walk.

Following the demonstration I let my victim sit down without any further explanation. Then about halfway through the meeting I pulled the step back out and asked my "volunteer" to join me back at the front. To be fair I had told them each that they would be playing my game, I just hadn't told them what that meant. The second instruction was this:

> Please step on the step again using one leg at a time. Begin with your right leg and use only the muscles in that leg both to step up and to step down. Then switch and do the same with your left.

Most of my participants were bewildered (as intended) by my instructions. Some tried and failed. Some worked it out in real time and clunkily got it done. One asked me to show him how.

The point, albeit an incredibly simple one, is that there is a very distinct difference between the "idea" of being aware and the "action" of being engaged in your process. So often supervisors, managers, safety practitioners – "leaders" – assume our people know how to recognize hazards and work through their own problems. Some do, but it's a dangerous assumption.

So, here's your tool. Give your people a plan to work through, not a task to execute. Teach them how to work through it and why the plan is laid out like it is. Then give them the expectation to stop if that plan has gaps.

13

BUILD A FOUNDATION, NOT A RULEBOOK

About a year after my fateful day with Inspector Bob, my company was in the process of figuring out how to correct a misstep that had led to some uncertainty and bad feelings regarding our safety culture. What they had done was well-intentioned, just not well thought out. It was a mistake that companies make all the time, almost maddeningly so. Maybe you've dealt with it yourself. Here's what went down.

In the interest of drawing a line in the sand, the Director of Safety had released five "Cardinal Rules." In theory, these were the five most important safety topics one might encounter on an average day at any of our facilities. There were two problems with these rules, however.

The first problem was that the five topics really didn't make sense. One of them, for instance, was "Job Hazard Analysis" (JHA). Employees constantly questioned the rationale behind that rule because, well, there wasn't any. It wasn't a high risk task or subject such as the others (confined space, lockout/tagout, etc.). Important, yes, but it didn't make sense in the context of a "Cardinal Rule." What would happen if you didn't complete a JHA? What would happen if you *did* complete one, but did it poorly? Did every job need a JHA? No one knew.

And that brings me to the second problem: There was no foundation behind any of the rules and even less framework around their enforcement. Employees were unsure of the expectations. They knew that the term "Cardinal Rule" sounded ominous, and assumed (sometimes correctly) that it was associated with discipline, even termination if one did not comply. But no one knew what being in compliance meant.

This had been a mess from before I had even arrived at the company. Even after over a year of backtracking we were still fighting to recover from that one poor choice. There had been attempts to make these edicts impactful, but most everything that had been done up to that point had only added to the unrest. A supervisor, for example, had received a final warning in his file for identifying that he had accidentally broken the plane of a confined space (put part of his body into the opening) (no one saw him, mind you, so he could have just kept his mouth shut ... so much for trying to do the right thing). At the same time, a technician made nearly the same mistake and was only verbally counseled. This was done to show that there would be no favoritism based on your level within the organization, but it really only served to add to the confusion. We were living in a culture of fear where employees were not concerned with doing what was right because it would keep them from harm but were instead operating solely with the intention of not violating a rule.

I don't really have any life-altering moment of learning or fun escapade to share in this chapter, but I'm including this topic because it is a vitally important part of successfully managing a workplace safety program. I'm sure many reading this have struggled with the exact (or at least dramatically similar) scenario as the one I just described. Too many companies, even highly successful ones with "great" injury rates, simply regurgitate OSHA regulations onto a few hundred pages of paper, slap their company logo on the front, and call it their safety program. That's not so much an indictment as it is a simple statement of fact. Those who have traveled around this country (or others I'm sure, although I'm no expert) know that maybe only one in ten employers has a truly solid safety program. A program that everyone understands, uses, and sees benefit from. Sure everyone talks the talk, but the disconnect between a book with a bunch of rules and the worker expected to live by them is huge.

Let me illustrate the point this way. Let's say you go out onto your site tomorrow and find a worker who is "violating" a safety rule. We'll assume that you use Toolkit Item 4 to be useful, so the worker is not on the defense. At this point you haven't addressed his or her safety infraction, so consider performing the following exercise before you initiate that part of the conversation: Ask yourself if you can be certain that the worker has ever been taught the correct way? If the answer is no, there is no need to continue. Instruct the worker how to complete the task safely, and then get to work determining where your program is lacking. If the answer is yes, your mission is now to figure out why in his or her mind the risk of performing that task unsafely was worth taking. Maybe it's production, maybe it's lack of understanding, maybe you just ran into the one person who just doesn't care – there are a million and one possibilities. Whatever the cause, the fact remains the same, your program lacks a strong enough foundation needed to support its objective.

When we realized this about our program, we actually didn't have any clue how to remedy the problem. As I've already described, we were essentially trying to build our battle plans while already taking enemy fire. Many organizations are in that predicament. The ones who successfully make it to the other side are the ones who know you have to eat elephants one bite at a time but aren't too prideful to invite the rest of the neighborhood over to share the BBQ. So that's what we did.

First, we listed what we wanted our program to include. We studied examples of safety programs from similar companies, read trade publications, and solicited feedback from anyone willing to provide it (there were many). Once we picked our "table of contents" we got to work developing programs. The true success of that phase wasn't merely the process of putting ink to paper and then making everyone in the company sign off stating they would comply, it was in promoting ownership. Armed with our list, we reached out to the actual people who would be using these programs and asked them for their help. We started each new standard, procedure, or guideline with two requirements: (1) it had to meet the law (federal, state, local, company, etc.); (2) it had to be operable in real life. Then we got to work.

In the end, we executed a very successful rollout of new processes and training which reached far beyond a few arbitrary rules. It was a long, grueling process, but by focusing on building programs people could and *wanted* to use, we managed to shift the focus from compliance to cooperation. We kept the "Cardinal Rules" in spite of the damage they had caused, but designed training and information campaigns aimed at demonstrating the potential impact each of those subjects could have on a person's life. They were no longer threats of the punishment that would be imposed if a rule was broken. I would actually argue that organizations who build their foundations properly don't need "Cardinal Rules," but that's another discussion for another book.

There are plenty of ways to slice this pie, and by no means do I have all of the answers. Maybe your organization has done some remarkable things that you'd be willing to share. If you have your own successes, share them on social media using the hashtags **#safetyfoundations** and **#relentlesssafety**.

Stupid Simple Toolkit Item 8: Never Assume Prerequisite Knowledge

I vividly remember my professor discussing Erikson's Stages of Psychosocial Development during Psychology 101 in my freshman year in college. I'm not going to try to make myself sound more educated than I really am here, but one point he made stuck out not so much for its academic significance, but for its anecdotal similarity to the safety profession. He was discussing the transitions from the 4th (Industry vs. Inferiority) and 5th (Identity vs. Role Confusion) stages of development and made the point that adolescents believe others are far more interested in their actions than they really are. I could picture myself at that stage in life during countless times where I had obsessed over my choice of words or the "awkward" way I had spoken to a cute girl, mistakenly thinking the others involved in those interactions had even given my words more than a fraction of a second's thought. The professor made the point that it is critical for adolescents to come to the realization that no one really cares that much about what others do or say. I say all of that just to joke that most safety managers never develop past that awkward teen phase of insecurity and self-importance. It might not actually be a joke though.

Safety practitioners (actually any person in a leadership role) often fall into the trap of thinking that others know, understand, and value our chosen craft as much as we do. That's just not true. And it shouldn't be, because it's not the job of a welder, or carpenter, or (fill in the blank) to develop methods to manage risk. It's their job to "build and do" not "think and theorize." If we are willing to admit that, we can bridge the gap. We have to stop assuming that knowledge of safety procedures and requirements is a natural, intuitive part of life. It's not.

So here's your tool: use Toolkit Item 3 and write better to build useable foundation programs. Then train the crap out of them. Train until safety is muscle memory, not a reaction.

14

NEVER TRUST THE PRESIDENT

Abe Lincoln (you and I both know that wasn't his real name, but he really was named after a president so it fits) and I stood outside the main administration building of my region. We were still embroiled in righting the wrongs the cardinal rules (I'll refer to them as "bird" rules from now on) had wrought upon us, and Abe was expressing an ancillary concern. He was one of the company's highest executives and was trying to wrap his head around the apparent unwillingness people had for discussing safety concerns. It was not the first time I had been asked to postulate why reporting is not what someone believes it should be. More often than not this is due to a perception that safety-related incidents or issues should be reported at some specific, measurable rate. Quite honestly, I wish we could just get over the idea in general, but until safety practitioners quit shilling bunk safety science we only have ourselves to blame. Until then we'll have to continue dealing with the ghost of Herbert William Heinrich and his nonsense pyramid (if you're not familiar, Heinrich asserted in his 1931 book *Industrial Accident Prevention: A Scientific Approach* that there is a statistical correlation between minor incidents such as near misses, minor injuries, severe injuries, and so forth).

I doubt the notion will be going away any time soon, but "Heinrich's triangle" has been widely discredited in modern safety management models and there's plenty of credible work out there that has definitively debunked it. Study the work of Fred A. Manuele if you're interested in learning more about it. In any case, Abe had been fed enough fake safety science and had it in his mind that if we just upped the reporting of "near-misses," our incidents would decrease. To that end he asked for my advice on how to make that happen. I explained the following points to him as I formed my response.

1. *Stuff is happening and you're not hearing about it.* Leaders always seem to have a tough time accepting that they don't know everything about what happens under their command, but it's one of those bitter pills that have to be swallowed in order to learn and get better. It's actually pretty reasonable to accept this fact when you consider that people are hard-wired to not want their "dirty laundry" aired. The problem itself is the perception that talking about the state of safety is dirty laundry. We lose the ability to learn when our managers are conditioned to believe that. It's true that the consequences of an incident are quite often negative, but we tend to get hung up on that rather than just taking a deep breath and looking forward. If someone reports a hazard (near-miss or otherwise), it should not be considered indicative of performance (good or poor), only an assessment of the way things are at a given moment in time.

2. *Employees don't perceive any personal value (or may only perceive negative value) in reporting.* Whether perceived or real, culture often leads employees to believe they will be punished in some way if they report a safety issue. This condition often exists because management is prone to trying to fix an employee before asking how the system could be improved to prevent the employee from being exposed to the hazard in the first place.

3. *Employees need to recognize there is something worth reporting.* In a perfect world if a hazard is recognized (say someone walks through a space and notices a trip hazard) they stop and take care of it. Then, that employee casually and freely (without fear of any "negative" consequence) mentions it to a superior who, in turn, communicates it to everyone. The problem in our circumstance at that time, however, was twofold: (1) our foundation wasn't strong enough to promote the type of awareness needed to recognize hazards; and (2) talking about those things would get you slapped around and chastised. Or so everyone believed.

Up to this point Abe was nodding along with me in agreement, if even with the expected amount of skepticism. He realized I hadn't actually provided any direction and knew that I was probably setting him up for the punchline. My reasoning for providing that type of set-up was strategic, though. I was about to tell him something he *did not* want to hear.

"So how do we fix the problem, then?" he asked. I took a deep breath and considered my words carefully.

"We don't," I replied. "*You* do." If you haven't noticed by now, I have a habit of saying risky things to powerful people. That probably won't change any time soon. He eyed me carefully but let me continue without objection.

"The way you get people to talk is by standing up in front of everyone and telling them in no uncertain terms that any mistake will be forgiven, even if it meant you violated a safety rule." I continued on and explained that what I was talking about were those episodes similar to the supervisor who "self-reported" his confined space violation. Honest mistakes. Not habitual screw-ups and certainly not willful disregard for anyone's safety. Following that explanation, we sat in awkward silence for what seemed like hours. Finally, he spoke up.

"I can live with that," he said finally. He'd taken my bait and I knew at that moment I only had one chance to sink my hook.

"Are you sure about that?" I said, catching him off guard. "What if I make a mistake and someone dies? Will you forgive that?"

His answer to that question, as you might imagine, was an emphatic no. We didn't want "people like that" working in our plants. To be honest I wasn't expecting a different response, but I hope I planted a seed. Even if Abe never understood that actions and consequences are often mutually exclusive things, maybe some other executive who reads these words will get it. I'm not trying to diminish the severity of a worker death by any means, but my point could not have been more simple. Until we start addressing risks independent of our emotional response to the consequences they cause, we will not be able to objectively look at our organizations and make decisions that have real impact and provide measurable improvement.

Stupid Simple Tool Kit Item 9: Forgive Mistakes

This tool could not be more basic and should need little explanation. But using it successfully requires putting all of the other tools we've discussed previously into practice first. I won't summarize them all, because I think it's probably worth taking your own look back through them. One thing I want to be clear about here is that I'm not advocating for or against any type of discipline program or scheme that you should or shouldn't employ in relation to your safety program. That is a Pandora's box that I'm not sure I'll ever attempt to open. Enough babble, here's the tool.

Assuming you've built a foundation, become a useful resource, learned how to write, and so on, now figure out how to separate human emotion from your safety program. It sounds easy, but this is one of those pitfalls we're never going to become immune to – because most of us are human. This is going to mean something different in every organization, but there's a simple test to determine if you're using this tool effectively: Whenever you make a decision regarding safety performance, ask yourself honestly if you would have done it differently if the consequence had been different. If the answer is yes, you're not there yet.

15

THE PILOT'S PERSPECTIVE

Nick and I were standing at the window near our corner-office cubicles which overlooked the main drag of the iconic city in which we worked (use your imagination). He had just delivered some incredible news and was grinning his typical, smirky grin. I had just received a promotion and an unheard of 8% raise to go along with it. He waited eagerly in anticipation of my response.

I was beyond angry! It was one of the few times in my career I remember feeling as if I came unglued. I had missed the whole point about the raise and the promotion, and chosen to focus on the one aspect that felt wrong. My promotion meant I was now officially a "house-mouse." I had developed a program that the company was looking to make international and my time was to be dedicated to figuring out how to make it work. I had been snubbed of my hard-earned right (at least in my mind) to be one of the field guys. It was something I'd been working toward since the day I set foot on that site and that prospect had been snatched away from me in an instant.

I realize that getting angry about a promotion is a ridiculous notion. I realized it as it was happening. But this was one of those times in life when I believed I knew better than all those who had come before me. I just knew their decision was the wrong one. In truth, I couldn't see the forest through the trees at that moment and know now that if I hadn't gotten that "assignment," I wouldn't have any of the knowledge I'm sharing in this book. Nick let me rant just long enough for my ears to get their signature red glow and then stopped me dead in my tracks.

"Son, you need to fly the plane you've got!" he said more sternly than usual. Nick had been a pilot in the Air Force and flew for the CIA in Laos and Cambodia before the Vietnam

War. Every story he told had a plane in it, but this time the reference felt out of place and made absolutely no sense to me. It worked exactly as he had planned, though, and my confusion stopped my whining mid-sentence. I stared blankly at him and probably cocked my head to the side like a bewildered puppy.

"I have no idea what that means."

"It means you've got something people want, and you don't even know how to use it," he replied.

More confusion. If ever there were a sideways example that supposedly led to a point about something, this was it. Nick had a way of teaching through punchlines (obviously a style I learned to mimic). Usually, they came quickly at the end of a quippy anecdote or a joke about someone he'd known when he worked in China. I could tell this punchline wasn't coming quickly at all. He was transporting himself back in time, gearing up for a story that would take a little bit of a journey to play out.

"Do you know what happened in 1959?" Nick asked. I shook my head.

"You know I don't, Nick. I was negative 21 then." We'd had similar conversations previously, so don't be too impressed by my on-the-spot math skills.

"Well, if you remember," he mused, "I was a 19-year-old kid in Laos who thought I was hot shit."

"Yep." I tried to internalize my eye roll.

"Do you remember what I was doing over there?" he asked.

"Running drugs for the Viet Cong?" I tried to get a rise out of him. It worked.

"Bullshit!" he said. "You know everything in that stupid Mel Gibson movie," (*Air America* in case you're curious) "is bullshit!"

"OK, Nick," I relented. "What were you doing?"

He grinned. "We were flying that piece of shit F-117," he said. "Day in, day out. For no goddamn reason!"

"You got me," I said. "What was the reason?"

"Goddammit, I just told you. There wasn't any reason!" I nodded in ignorant agreement. "But do you know what happened the first time I got in that plane."

"Nope."

"I tried to redesign it!" That statement caught me off guard. I was expecting him to tell me something about how in that moment he had defeated the system or got one over on the man, but he was actually telling me the opposite. He described that cockpit in detail (something I won't try to recreate) and then explained all of the ways his 19-year-old brain didn't agree with the way it had been laid out. He mimicked all of his whiney, disapproving discontent and then described the moment when his instructor co-pilot stopped him mid-sentence and simply told him: "Son. Fly the damn plane you've got! Maybe one day you'll engineer something perfect, but today isn't that day. Grab that stick, and get us airborne!"

That story floored me. I have no idea how much of it was real or just an embellished memory. But I do know he was there, in 1959. And I know the plane sucked because I've sat in the very same type of cockpit. I hope my retelling of this story does it justice. Because the point of the matter was not that the plane was poorly designed or that the young upstart knew better than those who had come before. It was the fact that the world is what it is. You can certainly make it better in the future, but not without understanding where you are today.

The rest of the conversation we had that day will forever hold a place in memory, but it's between me and Nick. I'll most likely be sitting in a retirement home at 90+ years old bellowing unintelligibly about flying planes and being too tall to fit in the cockpit (if you remember how tall I am in real life, that's actually a really funny thought). But the lesson will never go without having been learned. Life throws us circumstances. You can learn to live with them and fly that plane like Maverick in *Top Gun* or you can play the victim and pretend that the odds will forever be stacked against you. It's your choice, but the outcome is never set in stone.

Stupid Simple Toolkit Item 10: Fly the Plane You've Got

This tool encompasses two things. The first are those things which you can control, the second are those which you cannot. Both are vitally important to understanding what we're working with here.

First, it would be easy to simply accept the status quo and resign ourselves to the perpetual nothing that those who have come before us have accomplished. That's the easy way out. It's the educated man's way of saying "Well, if they had wanted us to fly a better plane, they would have built it already." That's crap.

But the real visionaries already know what I'm about to say. They know that despite the odds being stacked against them, they can make something of a junk design and fly it as long as it's got wings attached. That is the kind of safety practitioners the world needs. We need people who are willing to look at what we've got, recognize its shortcomings, and fly it anyway. They do it even though they know it can barely take off the runway, but they grab the stick and get it airborne. You may not be in the ideal environment to make the kind of dynamic culture shifts this book encourages, but there is always something you can do. You can either sit on the runway and complain about how bad your plane is, or you can learn to fly what you've got. It's your choice.

I know this tool seems much more rah-rah, self-help than the rest of them, but it's here for a reason. If you've come this far, the mission gets much harder from here on out. Only the most resolute will continue and when it's all said and done your plane may have more bullet holes than you've intended. But it's time to get airborne.

So take what you've got, nose down the runway, and throttle back. We're not stopping now until we get this thing off the ground.

16

THE GECKO

"Oh Look! It's the insurance lizard." The melodic cockney-accented words pierced my silence and I immediately realized the visit was afoot. I had to take a deep breath and physically chomp down on my tongue to hold back laughter. The desktop background on the monitor of my computer proudly displayed one of the first memes I had ever collected which depicted a small lizard biting someone's finger and the words: "INSURANCE COLLECTION AGENT." I turned around in my chair, red-faced to meet my visitor.

The short (as in shorter than me) British man standing behind me was our division safety manager, straight off the plane from London. I hadn't expected to see him for another few hours. Secretly the team had referred to him as "The Gecko" for going on three years at that point. Partly because he talked like the one from all those commercials, but mostly because he never caught onto the joke and that made it so much funnier. We'd even gotten outright overt with our references, but he was either blissfully ignorant or a really good actor. Tony had a squishy hand exerciser with the logo on his desk, I had my desktop picture, Nick had a "save 15%" bumper sticker, even Kelly had played along and had a stack of fake money with googly eyes sitting on the corner of his desk.

The nickname was the only enjoyment associated with a visit from The Gecko. He was (probably still is) an insufferable man. One could genuinely, actively try to find common ground with him, and I often did, only to be passive-aggressively insulted away. One might, for instance, ask him about his early career upbringing as a "chippy" (carpenter) only to hear a response like:

"Oy, mate. You wouldn't understand since ya never put in that kind of hard work." Hypothetically speaking of course.

Interpersonal problems aside, though, he was the boss and we did what we could to appease. Even when he questioned *everything*. Since I'm on the subject, and because it would benefit the scientific process, I might as well continue sharing. One of his favorite things to do was to question the findings of our accident investigations. He was the best armchair quarterback I have ever met, even to this day. It was one of the reasons my accident reports were so dry and matter of fact, I wanted to leave no opportunity for second-guessing. But I was rarely successful in that.

One incident, in particular, has stuck in my memory for some reason above so many others that should probably have retained a place of prominence. The event was simple, cut, and dried: *An electrician was in the process of pulling a run of wire through a new conduit and fell down when the string he had tied on the end of his fish tape had sheared, causing him to lose his balance. The worker fell backward and struck the back of his hard hat* (which he was wearing properly, even though The Gecko couldn't be convinced) *against the concrete floor behind him. He sustained a small laceration on the back of his head and was diagnosed with a concussion from striking his head against the inside of his hard hat.* Simple.

The Gecko was never satisfied with simple though. He couldn't understand how it was possible for someone to fall from the same level and strike the back of his head and he was downright flabbergasted by the notion that the man's hard hat had allowed him to sustain a concussion. We went back and forth for weeks about this particular incident and he could not be swayed from disbelief. I don't know what it was about that particular injury, but he wouldn't let it go. He would send a line of questioning about the general housekeeping in that area which would prompt us to send pictures of a clean, smooth concrete floor. When that wasn't enough he would inquire about the tensile strength of the string or the amount of force the gentleman had used to pull against it. When those inquiries wouldn't satisfy he began making up hypothetical scenarios and conspiracy theories about elevated platforms that had been quickly removed before anyone could see what *really happened*. At one point he even hinted that he strongly suspected the man had been assaulted.

As I reflect back on that incident it serves as just another reminder about the dangers of judging your performance based on the consequences of past events. If he had been able to look objectively, The Gecko might have been able to realize that there were bigger fish to fry and greater risks to resolve. But as it was we were stuck running in circles trying to prevent something that had already happened. I'm sure more than a little of his motivation was aimed at trying to prove that the incident didn't really need to be reflected in our accident rate (and subsequently, his "poor" performance), but enough with that dead horse.

The Gecko finally broke me down. I had no more rational explanations and I didn't know how to make the problem go away. So, I did what I often do and wrote down what I *wanted* to say. When Nick read my response he chuckled and *almost* let me send it. Almost.

Pursuant to your request for updated information, we would like to relay the following:

The injured party is still in recovery but remains in stable condition. The injuries sustained correspond with the details provided by the field inspection team. Though not concluded, the investigation report will most likely confirm those details. At this time we have no reason to suspect that the employee was injured by any other circumstance than a fall from a level surface.

However, in your inquiry, you suggested that we further investigate the circumstances surrounding the incident to ensure that the fall was not experienced from an elevated platform. We have deduced thus far that while unlikely, the possibility of an elevated descent remains plausible. You will note from the photographs provided that there are no elevated platforms in the immediate area. This observation cannot be positively confirmed, however, due to the strong belief that an extraterrestrial life form had rendered any such surfaces invisible with a photographic cloaking device. We suspect that the life form may have been in the immediate area at or near the time of the employee's injury.

Additionally, in response to your query of a possible assault, we believe that this is also plausible. The same ETL (extraterrestrial life form) may have inadvertently caused harm (though not with malicious intent) to the employee when he activated his photographic cloaking device.

In no way are these theories considered fact. They should, however, be considered in the overall evaluation of the incident in question. Due to governmental restrictions, we will disavow any knowledge of this correspondence in future exchanges. Please do not reply to these comments.

Believe it or not, those were actual words that I typed, spell-checked, and then saved to a flash drive back in 2010. I know none of it was helpful, but it made me feel better. In the end, that episode came to a close, but I described it here to give you an idea what we were up against. One lesson I had not learned at that point in my career is that many times you can't win with unreasonable people. Sometimes you have to just take them for what they are and beat them at their own game. The Gecko was motivated by one thing: looking good. His definition of good was a fluid thing, but as long as he was getting his bonuses and not having to answer hard questions from his boss, he was happy. I would have been much happier at that time if I had realized that.

The day he happened upon my lizard picture, he had come to get a report on an experiment we had been conducting. It was not going well.

That moment I described in Chapter 15 ... you know, the one where I was angry about having to stay cooped up indoors? This is the culmination of that "program" I had written. I'm not going to lay it all out in graphic detail (mostly because it's boring as hell), but the long and short of it is that we had been playing a high stakes game of corporate poker since that moment and The Gecko had come to cash in his chips. Obviously, I'm not going to leave you completely in the dark so let me fill in enough details to make sense of it.

The program was a simple thing. It was based on the premise that if our people knew how to identify hazards, and were held accountable to eliminate them, we could improve safety conditions on the site. That premise is, and will forever be, infallible. The program was not though. We floundered with it for well over a year because we had failed to figure out what our real goal was. Let me back up.

When I got my big "promotion" and was charged with creating something meaningful, something that had an impact, I naively believed that my playing field was level and wide open. I could not have been more wrong. At the time, actually the entire time I had been employed with the company up to that point, I had been in charge of our project's safety "observation" program. It was a really messed up mish-mash of behavioral and condition-based mini site audits which were designed to "predict" where there were opportunities for workers to experience an injury. In actuality, all it really did was empirically prove that 13% of the time the observations were completely useless, our method for calculating "compliance" was positively skewed by nearly 8% to yield a higher number, and the observations rarely fit the scenario. (In case you're wondering, we really did study those data in detail. Those were the actual numbers we uncovered during an intense six-month long, six sigma performance improvement investigation.) All of that meant findings could only provide a proverbial "blip" on our safety radar. All of that amounted to indicating something might, maybe, and/or could be a risk worth looking into ... or not. Who knows? The point is that the program was an administrative, unjustifiably expensive waste of time. People like The Gecko were well aware of that fact, but at that point so much had been invested that no one wanted to muddy the waters. Except for some nobody upstart named ... Me. It was me. I can't help myself.

One thing I did have going for me was an unstoppable sense of optimism though. My goal wasn't simply to point out how much the process sucked. I really wanted to make a difference. So I set out, Nick in my corner, intent on paving a new road.

First, we analyzed the shortcomings of the current method: "Inspectors" were given a predetermined list (a checklist) of items to look for. Using the checklist, they were then asked to identify what was "compliant," what was "non-compliant," and what was "n/a." Those of you who are data-driven types are now probably doing the math to figure how I came up with the "usefulness" statistics I listed above (here's a hint: there were over 17,000 data points). Inspectors were also given a set of parameters that were intended to reflect a given project's success. Those were as follows:

1. Anything below 95% compliance was poor performance.
2. The Top 5 had to be evaluated even if they were not present on your site.

(OK that seems like it demands some explanation, but it really was exactly what it sounds like. You were *required* to perform observations on items that *did not* exist on your site. Only because they were determined to be a "Top 5" risk. Think about that for a minute. How valuable do you imagine an excavation observation might be on a site that has no excavations?)

If you think those are some pretty stupid rules to have to play by, well you're not wrong. But that was what our safety performance was being judged against. In this paradigm, surprisingly, OSHA Recordable rates weren't even a factor (mostly because they were in the tank). But what if your project had poor "prediction" observation metrics? Well, those could end your career. I know because they almost ended mine.

The catalyst for changing the program wasn't actually my idea. I'd love to say that it was, but it actually happened by chance one Saturday afternoon following a state-mandated training course. Following a new OSHA rule, all supervisory personnel were expected to participate in at least ten hours of safety familiarization training. Our team conducted the training dutifully and I'd like to believe that we actually taught some meaningful things to our students. The conversation that occurred below gives me hope that we accomplished that goal.

I was walking alongside our project manager (the same one to whom I had complained about Jerry's actions years earlier) on our site, and he was appalled by the general conditions he observed. We had been having struggles with the contractors at the time, so none of it was a surprise to me. Our manager, however, was not used to seeing poor housekeeping, unprotected excavations, or basic disregard for worker safety. But this wasn't a typical job for our company.

"Jason," he asked bewildered. "What was our prediction compliance score on housekeeping last month?" I looked over at him sensing the undertone.

"98% compliant," I said. Silence.

"How is that possible? This place is a mess." I knew the question was rhetorical, but again … I can't help myself sometimes.

"Because that's what you asked for," I replied. I think our conversation ended after that, but I do know that at some point he asked the safety team to figure out how it could be done better. Hence, "the program" was born.

Based on that request we built the first "stupid simple" tool in the kit. We simply asked people to find the things that could harm, fix them, and then go out and do it again. That's it. Within the span of one year, that initiative contributed (I'll never try to claim it was the sole cause) to a 50% decrease in injuries on that site. But the program was a terrible failure ... at first.

It failed initially because we didn't know how to keep score. See, we kept with the same terminology that had been our downfall from the beginning. Managers were conditioned to believe that "non-compliant" statistics were bad. Now, suddenly, they became inundated with metrics from our project which they had been taught to identify as indicative of *poor performance*. Overnight a project that had boasted 98% compliance in housekeeping month after month began reporting that our housekeeping conditions were closer to 70% ... or worse.

But we missed the point when we reported those numbers. We missed a key opportunity to capitalize on the actions we took each and every day to remedy those hazards we identified. That was the real metric. Everything else was just a snapshot, a picture of the way things were at a given moment in time.

The day he showed up unannounced, The Gecko had been expecting our apparent, inevitable defeat. What he didn't know is that we had figured it out and had beaten him at his own numbers game. We may have only had 70% "compliance" with having a non-hazardous worksite. But at the end of that day, we proved that we could eliminate those hazards nearly 99% of the time as soon as they were identified. Even skeptics would agree there's value in that.

Stupid Simple Toolkit Item 11: Measure the Right Things

Author Sharon Gannon said "We create the world we live in. If we want to change what we don't like in the world, we must start by changing what we don't like about ourselves." That's a concept I hope I have not just lazily glossed over in these pages. The state of the industry and the safety of the workers who are in it is the result of what its leaders do. That includes everyone who decides to put the word "safety" into whatever professional title they hold. While the concept certainly means that we can (and should) work our hardest to make an impact, it's often a stark reflection of our ineffectiveness. I don't say that to chastise or debate, that's not the point of this book. My goal from the start has been to provide a hard, honest look in the mirror and then provide something useful to help move the dial in the right direction. Even if it only moves in tiny notches.

The fact of the matter is that safety performance is measured the way it is because that's what we have been telling our leaders to do for decades. We need to pull their notes, recalibrate their thinking, and write them a new script. One that doesn't just placate, but resonates with workers and sets realistic goals everyone can work toward.

Fair warning, this tool takes buy-in and commitment from more than just "The Safety Guy." It will require some real faith from your leaders and it will feel like scuba diving without a mask for a while. Well, suck it up and practice holding your breath. The tool is this: Stop measuring performance with rates; start measuring action. Never mention an incident rate again (to an employee). That's it. Simple right? I'll give some suggestions on how to accomplish this in Part III. Don't worry, it's not as scary as it sounds. And if you are worried about it, don't be. Your incident rate won't go away. It will still be there to coddle and console you in moments of doubt and weakness. You're just not going to talk about it ever again.

PART III

DO WHAT MAKES A DIFFERENCE

After my sophomore year in high school, I decided that I didn't want to *do* high school anymore. It was something I whined about frequently, particularly to my parents. I learned quickly that whining typically falls on deaf ears, though. My mom had finished high school, my dad had finished high school, and I *needed* to finish too. And then a stroke of genius hit me.

This was before the days of Google, so I really don't remember how I figured it out (who knows, maybe I went to a library), but somehow I learned that the State of California would let you "test out" of high school. It was (still is) called the California High School Proficiency Exam. If I were eligible for the program, I would be able to take the test and then be granted the legal (at least in the State of California) equivalent of a high school diploma. It was *not* a GED (even though everyone I tell this story to replies back with "Oh, so you got your GED?"). So, I petitioned my parents to let me apply for the test and then attend community college instead of returning for my junior year of high school. They appreciated the initiative and agreed. But there was one minor roadblock.

The application for the test requires a school official's certification and signature. That meant my dad and I had to make an appointment to go talk to Principal Doug. We did, and I'm sure that as I told my story I came off as every bit of the cocky 17-year-old know-it-all I was. It probably sounded more like a diatribe on how terrible his school was than a plea for self-improvement. In the end, he signed the form, but not before mentioning something that has stuck with me since.

As we were leaving, Principle Doug said, "I just don't think you should quit high school. I mean, just because I don't like my job doesn't mean I can just quit."

I looked back at him and without missing a beat asked, "Why not?" Doug was not expecting me to fire back like that, but I could see my dad grinning in the corner waiting to see what would happen next.

"Well, because," he considered his words carefully. "It would be hard."

He was not wrong about that last point. But as I've already covered, the fact that something is hard to accomplish is no reason to be dissuaded. Principle Doug lit a fire of resilience that day. I needed to prove I could do something no one believed in. This book, more specifically the things I'm going to cover in the next three chapters, are a continuation of that attitude. The concepts here are no different than what we've been slogging through over the past pages. They're so simple, they sound dumb. I am not going to try to reinvent the wheel or even say I've come up with some new and innovative way to use an old trick. But anyone who's spent any time in this field at all knows that STUPID SIMPLE ≠ SUPER EASY … or easy at all.

I'm taking the time to write this out because it's easy to take the simple things for granted. One lesson in leadership that I've carried with me since the military is the idea that you don't overlook the small stuff because that's what gets you killed. Sometimes you have to spell out every detail of every expect-ation, and then repeat yourself again and again. That concept is lost in industrial safety because we (wrongfully) assume safety is understood and valued universally. If we can get past that misconception, we might actually have a chance to make some-thing where there wasn't anything before. Then we have a real chance at success. The next three chapters are what I believe and have seen can drive real change and help grow a culture that is a self-perpetuating example of **#relentlesssafety**.

17

WE SHOULD DO SOMETHING!

The words were highlighted in bold italics in the latest email from our Director. My counterparts and I were all sitting in different places across the country, but we were all scratching our heads trying to figure out what "something" meant. We had less than one day to figure that out and reply with our response. I won't leave you hanging too long. We got it wrong.

It was June 16, and the message read as follows:

> Guys, as you know, June is National Safety Month. The plants have a keen interest in this event, so *we should DO something!* I would have expected one of you to bring this up sooner since you are the leaders of the Company's safety program, but this was brought to my attention by others who do not want to pass up this opportunity.

Did you catch the tone? Oh that's right, emails don't have tone. Disregard that question. Maybe you did happen to notice the date though? June, freaking 16th! That's when she decides to tell us we suck? She was enjoying a Mediterranean vacation and hadn't been in email contact for two weeks mind you, but apparently this was important enough to interrupt the sunbathing. So ... we did ... something.

The something was exactly what anyone who was cornered, insulted, and not just a little aggravated would do. We performed an emergency R&D (Rip-off & Duplicate) operation. We did actually know that June was the National Safety Council (NSC)'s National Safety Month, but we had made a collective decision earlier that year to focus our initiatives on issues that were more industry-specific. She didn't remember, but that didn't matter. So we hopped onto the NSC's site, shamelessly copied their (free) weekly safety topics for the month, and fired back at her.

The reply email was much worse. There was so little that was constructive or even entertaining about it that I'm not going to give it the time of day here. Let's just say she was less than impressed. I imagine some of the strife was caused by the fact that we were all embroiled in fixing her mistake from years earlier when enacting the "Bird Rules," but that's neither here nor there. The crazy part was that in the end she just sent our R&D'd material on and passed it off as her own. Even after rejecting it outright.

I tell that story, not only as another example of where safety can go off the rails and get caught up in things that don't matter, but mostly because it exemplifies what I'm going to cover in this chapter. How often do you, or a supervisor, or any leader for that matter look people squarely in the face and say, "Hey, go do *something*." You may not use those exact words but without clear direction the message is 100% garbled with all of the noise, all of the assumptions, and all of the good intentions that pave the road to, well you know. We do it to our people all the time. Just think about every "new hire orientation" where people were flogged with eight hours of PowerPoint slides and then deemed "safety trained" and ready for the field. Or what about the one-page lessons we print off and pass out in a morning meeting just to have a record that everyone was "briefed" on some new safety requirement. Too often we lie to ourselves and say that those initiatives coupled with some fancy posters on the wall mean everyone knows what's expected of them.

One perfect example of that was a recent time when I happened upon a worker who was clearly frustrated with something. I walked over and asked him what was going on. He proceeded to show me that the electrical shut off on a piece of equipment he was attempting to turn off and lock out was damaged and he couldn't get the machine to power down. I calmly suggested that he call the supervisor over and get it fixed when a panicked look appeared on his face. I'm not going to retell the story in conversation, but the long and short of it was that he had been willing (and if I had not shown up, probably would have) to take a chance and clean the machine without locking it out.

From the 30,000 foot view, someone who would put themselves at that type of risk doesn't make any sense. I mean, everyone knows that lockout is non-negotiable, right? Sure they do ... I mean maybe ... probably. But if something isn't negotiable, why do we continue to experience failures in those areas?

Because things get in the way of absolutes all the time. We're clouded and irrational human beings who are prone to mistakes, misjudgments, and just plain old ignorance. Sometimes it's because "production" comes first, sometimes it's because of some perceived deadline – there are infinite reasons. Those are issues that would take an entire other book to unpackage, and they're worth considering, but that's not what I'm getting at here.

The crazy part of that example I just gave is that the machine was a portable unit (albeit powered by an industrial 480-volt plug) which could be turned off upstream, unplugged, and replaced with one of three spare units that were sitting in the corner. The worker may have been testing my metal a bit too, but that's beside the point. What if your process, and the training that supported it, was such a staple that not only did people do it, but they didn't even *want* to question it? You might wonder if something like that exists or is even possible. Well, get ready for some more stupid simple.

When Padzilla and I were in tech school, I was assigned as the class leader. That meant that for every exercise or training activity we participated in I was responsible for assigning the crew to their individual tasks. It was a simple, rudimentary, repetitive process designed to drill procedures into our heads. And the best part about it? Safety was built in. Let me explain.

We were Munitions Technicians (AMMO as it's known in the Air Force). Our primary mission was to build, then transport munitions to our aircraft. We were also highly motivated not to explode ourselves or melt someone's face off by igniting an aircraft flare. But like I said, safety was built in. It wasn't done in the form of a "pre-job" brief or a Job Hazard Analysis (JHA), it was actually part of the job.

In training, one of the duties I was expected to assign each time we learned a new weapons system was (believe it or not) the role of Technical Order (T.O.; instruction book) reader. That person had no task other than to stand in front of the crew and read each step of the job from the T.O. aloud. And believe me the steps were *stupid simple*. They had to be, because if we're being honest with each other, the job of "bomb builder" is not often given to future brain scientists. I may know how to write words gooder than some people, but by no means am I the world's next Einstein.

The reader would stand at the front of the room and methodically take us through each step of the process. "Step 1 – using the green screwdriver, turn the middle screw one quarter turn counter-clockwise." He would then dutifully place his finger on the page where he left off and look up to verify that the quarter turn was indeed accomplished. "Step 2 – push the quick release button on the side panel next to the middle screw." Again he would look up, finger keeping tabs. The steps would continue until in big bold CAPS LOCK the words **WARNING, CAUTION,** or **NOTE** appeared. At that point, the reader would loudly read whichever of the three words were in front of him and then look up to the rest of the crew for acknowledgment. *"Warning,"* he would say. *"Warning!"* came our reply.

The instructors would actually make us stop and repeat the step if even one man on the crew missed the cue. Once we acknowledged, the reader would continue reciting a carefully constructed safety note that required absolute attention lest ye be blown up or at the very least lose a limb or two. As you might have guessed, *warnings* were items considered deadly (as in actual life and death) serious, *cautions* were things that could get you a trip to med and a liver-killing 800 mg Motrin habit, and *notes* were things that would likely mean a trip to the first aid kit and an annoying incident report to fill out.

We performed every operation, no matter if it was attaching a high-explosive warhead to a missile or changing a tire on a trailer, in the same fashion. There were never exceptions. To this day I still fight the urge to yell "Warning!" back any time someone uses it in a sentence.

Once we graduated training the process got quicker and we didn't always read aloud (although it was the go-to method any time we trained someone new or performed a task that hadn't been done in a while), but we didn't need to. The thought process was built-in and integral to everything we did. I'm not saying we never had injuries or near misses, but one thing I am certain about is that we always knew what we needed to control. It started from the moment each person was given a clear and understood task which came with clear and understood expectations about how to complete it. The best part about it was that on top of the knowledge we had been given to *do* the task, we were also armed with the respect of knowing that

"*WARNING: This WILL blow you to pieces if do it wrong!*"
That kind of thinking is non-negotiable.

A simple way to think about putting this concept into
practice in the industrial world is to ask someone to tell you
the first rule of safe lifting. Inevitably the person will say
"Lift with your legs, not with your back." When they do, ask
them how.

In response, most people will mime squatting down to pick
up an item. But pay attention to how they do it because most
adults will get it wrong (anatomically speaking that is). It's
a shortcut we learn over the years in our body's attempt to be
as efficient as possible. Efficient, not strong. Not "safe."

Most adults will assume a narrow stance and then squat
down onto the balls of their feet with their heels in the air. This
puts an intense amount of strain on the ankle, hip, and knee
joints and directs tremendous shear force through your patella
as you move your weight through space. A "real" squat should
be done flat-footed and balanced. If you're struggling to picture
what I'm saying, imagine a toddler squatting down to pick up
their heaviest toy. The child will instinctively do it right because
they haven't learned bad habits yet.

My point in saying all this is not to encourage everyone to
go get a personal trainer and learn how to squat correctly
(although I would never *discourage* it). It is to point out that
we take even the most basic of safety "requirements" for
granted and assume they are natural and intuitive actions. As
I've said countless times before, they are not. We need to recognize our own humanity and start teaching people all of the
things we assume they already know.

So let's say that you've been nodding along and you're committed to trying some of the things I've been advocating in this
book. Let's say that you start building a foundation by partner-
ing with the people who do the work and you build the best,
most awesome safety program the world has ever seen. I'm not
patronizing, I think that's a real possibility for some out there.
Now, what do you do with it?

Train. As the philosopher Archilochus said, "We don't rise
to the level of our expectations, we fall to the level of our training." It's a common mantra used by the military and I've even
heard it pretty frequently in the field of Emergency Management. But keep this chapter in mind when you train. Build in

your own warnings, cautions, and notes and make people yell them back to you until it's second nature. It seems stupid. It seems like it should go without saying. But when you think about it, the alternative is simply telling your people "Hey, we should do something."

18

"SAFETY ASSESSMENT STICK" BEATINGS

The first time I met The Gecko, he had just wrapped up a week's long audit of our site. This was months before Nick arrived, so Jerry had been my only (non-) buffer. As a gung-ho new hire, I had no preconceived notions about how to handle someone like him. I was actually kind of excited to learn from someone who had a different view of industrial safety from another country.

I should have realized that his first request of me was one of those "indicators" (they don't always have to be safety hazard related) I've mentioned previously in this book. He simply asked me to build him a "Dashboard Report" that summarized his audit findings. I took the order like the good Airman I still was, worked dutifully for several days, and then delivered it to him.

"Nah, that's not what I wanted mate," was his reply. We went back and forth like that a few more times before I finally made something passable in his eyes. It was yet another example of the clear communication subject I just discussed. Not that I blame him, all humans have the bad habit of assuming that everyone else knows what's going on inside our own heads. The "indicator" part was something that I wouldn't catch on to for a while though. The "Dashboard" he wanted was supposed to include all manner of pretty graphs and charts that sliced and diced his audit findings and pointed a glaring red finger (I actually made a chart that looked like a finger) at all of the site's deficiencies. There was nothing constructive in it, it was just a picture (as ugly a picture as I could draw him) of how poor our performance was. That picture was painted with hundreds of data points that tallied up each "infraction" The Gecko had identified. I remember him and Jerry having a little bit of a squabble about how unfair it was that he had marked us down for every ladder that was not properly stored in a rack rather than just making a general note and moving on past that topic.

One thing The Gecko did not believe is the notion that not all hazards are created equal. I guess you can argue that many little things could add up to something catastrophic, but there's no way to prove that. And even if you could, casting a broad net and making general statements about how bad (or good for that matter) a particular site is doesn't give you any actionable targets. It just gets people worked up and running around trying to do ... *something*. We're geared to aim at the easy targets and low hanging fruit so it may seem counter-intuitive, but often the best plan of attack is to take out your biggest obstacles first. The problem with that thinking is that safety practitioners aren't particularly good at identifying what the big targets are. As in which ones matter the most. We're great at finding things that are "non-compliant" or identifying regulatory (OSHA or otherwise) "violations," but we're downright terrible at prioritization. I call it "racking and stacking."

The result of our disorganization is what I have come to affectionately term "Safety Assessment Stick Beatings." I said it as a joke once (although my audience and I vehemently disagreed on the humor in it), but as I've progressed through my career I've noticed that even the most well-intentioned audit or assessment can be received as a beating when the message isn't delivered properly.

Some hazards (Francis's nail for example) are unlikely to ever be noticed, let alone cause harm. Take a deep breath and accept that; it's reality. Other hazards will kill without warning and need immediate action. We need a laser focus that guides our decisions and removes those risks first. Go ahead and knock out all those loose nails when you get a chance, but the real mission is keeping our people alive with their limbs attached, not saving money on ice packs and Band-Aids.

There is one more lesson to be learned from the WARNINGS, CAUTIONS, and NOTES I mentioned in the last chapter. It's nothing mind-altering. One of the most important things about those indicators was their clear, unchanging, descending order of magnitude. Warning always meant death. Caution always meant injury. And Note always meant ... well note, but something much less significant than the other two.

I'm going to challenge you to start looking at your safety program through that lens in this chapter. There are two paths to doing that. One is looking at the past and the other is planning for the future.

Anyone who dismissed the earlier parts of this book and put it down based on my zealous distaste for incident rates is going to miss out on the punchline I've been waiting to deliver since I started. You may recall my assertion that rates are not performance measures, but what I didn't address was their value (it's very minor so don't mistake what I'm saying here) as an indicator. They're not going away, so we may as well use them as intended, right? Here's how to do that.

Knowing that the criteria for determining rates are, well, stupid and arbitrary, we need a way to objectify them. There is no perfect method for doing this, but I'm going to share one way that has helped me. Feel free to tinker around and figure out what works for you. And keep in mind that there will always be some subjectivity to it. Practice will help. Consistency is paramount. Beyond those warm and fuzzy life lessons, there are three critical keys to achieving success with the practice I'm about to describe.

1. Your grading scale must be an odd number so you have a true middle – mine is from 3–13 (because 13 sounds ominous).
2. Your term definitions must be set in stone and easy to understand.
3. You must answer question four honestly. You'll see what I mean by that in a minute.

So here's the exercise. Every incident, event, safety-related thing that happens on your site gets graded (even if no one was injured). In the aftermath you answer four questions:

1. What actually happened?
2. What could have happened? As in, how much worse could it have been?
3. How likely is it to happen again (or does it already happen often)?
4. Is there a reasonable corrective action?

An example of how this could be done is shown in Figure 18.1.

Incident Severity Assessment

Criteria	Scale	Score
What was the _**actual**_ severity of the incident?	1= Near Miss ☐ 2= First Aid ☐ 3= Medical ☐ 4= Lost Time ☐	0
What was the _**potential**_ severity of the incident?	1= Near Miss ☐ 2= First Aid ☐ 3= Medical ☐ 4= Lost Time ☐	0
Has the incident ever, or is it likely to occur in multiple departments/sites?	1=Never ☐ 2=Seldom ☐ 3=Often ☐ 4=Frequently ☐	0
Is there a reasonable corrective action?	0=No ☐ 1=Yes ☐	0
	Severity	0
	Corrective Action Plan (CAP)	

KEY

7 or less = CAP Optional
8 or greater = CAP Required
Max score =13
Min score = 3

Hazard Severity Index			
3	4-6	7	8-10	11-13
Low	Medium	Med-High	High	Severe

Figure 18.1 Blank incident severity assesment worksheet (example)

As shown in the figure, each different answer to those questions will assign a numerical value. When totaled, you will have a score somewhere between 3 and 13. Then you simply rate that against the severity index and determine how much of a threat it is, thereby determining how much attention should be given to that event.

Doing this exercise may seem like a monotonous waste of time but, before you dismiss it, consider the potential benefits. First, you'll have a method which for the most part objectively determines what is and isn't a high-impact event. This will help you as you communicate with managers and leaders throughout your organization because it gives you the ability to remove the emotion from an injury simply because it was an OSHA Recordable, for instance. On the flip side, it will also help you shine some needed light on events that are incorrectly assumed to be on the lower end of the spectrum simply because the consequence was just a scraped knuckle or bruised elbow. Some of those are the most deadly.

In my example, everything above a score of seven would require some sort of plan to prevent a similar incident (I called it a CAP for brevity purposes). Any seven or below would be a discretionary choice, allowing you to "let it go" and get after the bigger fish. You might also notice that there's no mention of any catastrophic event like a death or dismemberment. It should go without saying, but those would be automatic 13s.

Also, that thing I said about question 4 ... think about it objectively. Everyone will likely have a differing opinion, but my suggestion is to discuss that question openly and honestly with your leadership team to find out what "reasonable" means to them. Try to remove the emotion and remind them of the goal (go back and look at Toolkit Item 2). The example I often give when referencing that question is an "incident" where an employee walks down the hallway, trips, and almost falls. He dutifully reports the near-miss and you go take a look at the area where it happened. There are no loose tiles, there's no furled up rug, no one spilled their morning coffee, you get my point. It's great that the employee reported it, and he should be praised for that. Maybe you even communicate it to the rest of the company. But if we're being honest, there's not much value in investing your organization's time and effort into an incident like that. People often think I'm callous and uncaring for saying that, but as I've already laid out, our business isn't removing life's bumps and bruises, it's making sure our process doesn't kill people. Too often we forget that. Figures 18.2 and 18.3 are examples of incidents that would fall on opposite ends of the spectrum I just described.

Now onto the second part. How do you use your Safety Assessment Stick to beat back risk rather than beat down your leaders? That's simple as well. You use the same method I described when looking at the past, except you remove question 1. Again, you can change it to fit your organization. But now rather than looking at a past event, we're grading hazards. One note of caution here, this line of questioning should be done independent of any implications of regulatory fines that could potentially be imposed. Fines may well be indicative of severity (though not always), but in this case since our objective is protecting people we want to avoid the impression that our motivation is in any way monetary. When you identify your hazard, violation, unsafe condition, and so on you simply ask the following:

1. How much harm could this cause?
2. How often does it occur?
3. Is there a reasonable corrective action?

Using that line of questioning, maybe your "rack and stack" scale now only goes from 2 to 9, like the example in Figure 18.4.

You may be thinking that what I've just described is nothing more than a typical risk assessment. You'd be correct. Nothing here is the magic safety bullet everyone's been searching for. But it is a practical and under-used method you can use to determine your priorities. Sure we all talk about it and learn how to plot things out, but how many of these basics have we allowed to slip away and just become talking points instead of useful tools we can use to educate and help guide the decisions of our leadership? Assuming *they* just inherently "get it" is just as bad as when we make that assumption about our workers. Maybe even worse.

What I'm suggesting here is that we start using what we have to objectively analyze what we should do. Then use that information wisely to help encourage sound business decisions.

Incident Severity Assessment

Criteria	Scale		Score
What was the *actual* severity of the incident?	1= Near Miss ☑ 2= First Aid ☐ 3= Medical ☐ 4= Lost Time ☐		1
What was the *potential* severity of the incident?	1= Near Miss ☐ 2= First Aid ☑ 3= Medical ☐ 4= Lost Time ☐		2
Has the incident ever, or is it likely to occur in multiple departments/sites?	1=Never ☐ 2=Seldom ☑ 3=Often ☐ 4=Frequently ☐		2
Is there a reasonable corrective action?	0=No ☐ 1=Yes ☑		1
	Severity		6
	Corrective Action Plan (CAP)		Optional

KEY

7 or less = CAP Optional
8 or greater = CAP Required
Max score =13
Min score = 3

Hazard Severity Index			
3	4-6	7	8-10	11-13
Low	Medium	Med-High	High	Severe

Figure 18.2 Completed incident severity assesment worksheet (low severity event)

Incident Severity Assessment

Criteria	Scale		Score
What was the *actual* severity of the incident?	1= Near Miss 2= First Aid 3= Medical 4= Lost Time	☐ ☐ ☒ ☐	3
What was the *potential* severity of the incident?	1= Near Miss 2= First Aid 3= Medical 4= Lost Time	☐ ☐ ☐ ☒	4
Has the incident ever, or is it likely to occur in multiple departments/sites?	1=Never 2=Seldom 3=Often 4=Frequently	☐ ☒ ☐ ☐	2
Is there a reasonable corrective action?	0=No 1=Yes	☐ ☒	1
	Severity		1C
	Corrective Action Plan (CAP)		Requi red

KEY

7 or less = CAP Optional
8 or greater = CAP Required
Max score =13
Min score = 3

Hazard Severity Index			
3	4-6	7	8-10	11-13
Low	Medium	Med-High	High	Severe

Figure 18.3 Completed incident severety assesment worksheet (high severity event)

Hazard Severity Assessment

Criteria	Scale	Score
How much harm could the hazard cause?	☐ 1= Near Miss ☐ 2= Equip. Damage ☐ 3= Injury ☐ 4= Death	0
How often is this hazard encountered?	☐ 1=Unlikely ☐ 2=Seldom ☐ 3=Often ☐ 4=Frequently	0
Is there a reasonable corrective action?	☐ 0=No ☐ 1=Yes	0
	Severity	0
	Corrective Action Plan (CAP)	Optional

KEY

5 or less = CAP Optional
6 or greater = CAP Required
Max score =9
Min score = 2

Hazard Severity Index			
1-2	3-4	5	6-7	8-9
Low	Medium	Med-High	High	Severe

Figure 18.4 Hazard severity assessment worksheet (example)

Think about it this way. If you charted every safety issue (regardless of the title anyone assigns it) similarly to what I've just described, you would be able to intelligently explain the success (or failure) of your safety program. It would provide you with a well defined, easy to understand measure that objectively shows progress and provides a target at which to aim. The alternative is continuing to use injury rates and then crossing your fingers and hoping no one strains a finger and has to wear a splint on it. Continuing down that path is equivalent to playing whack-a-mole blindfolded.

19

THE ICE METHOD

The "program" I've alluded to several times previously will always be one of my proudest accomplishments while always lingering in the back of my head as a continuous source of self-doubt. Part of me has always been tempted to believe it was too simple and that the improvement we witnessed as a result was mere coincidence. The evidence has always argued against my doubt, but if that were enough to satisfy me I wouldn't have gotten past the first chapter of this book. I want others to travel this road with me and see where it leads.

I've been pretty coy about how any of the mechanics of the program worked for good reason. Mostly because the drama that played out as we found our way really didn't have any influence on the outcome. Maybe I'll tell that story some day. I could describe the debates about behavior based observations versus the value of identifying hazardous conditions. I could go on for pages about how office politics threatened to trump worker safety. I could explain how a 70-year-old man put his faith and career in the hands of a 20-something nobody who had no business trying to change the world. But none of that will change what we discovered. None of the math, the analysis, the reports, nor the metrics were the keys to its success. The program's success was founded in its premise: *Attack the hazards.* That's it. You can slice and dice the numbers any way you want, but if you can't figure out what will get your people and then do something about it, your program is just a bunch of numbers. That is how we made a difference.

I.C.E. stood for Identify, Correct, (then) Engage. That acronym might seem really clever, but it was born more of a necessity to have a cool brand than it was to send a message. The crazy thing is that it became a battle cry for our project in spite of the cheesiness. Coincidental as it may have been, the idea actually makes a ton of sense when you apply it to real-life safety. It simply means that you IDENTIFY every hazard you see, then CORRECT it as quickly as possible, and ENGAGE your people both to educate them and to keep it from happening again. That is all we did, literally.

There were no compliance checklists, no "at-risk" behaviors to identify, no titles at all. There were just hazards. Problems that could harm and needed to be dealt with. We tracked every one until it was gone, and from that we witnessed magic happen. It wasn't mystical magic, it was human magic. Born of a desire to be the best (something I would argue lies in each of us). Every time one of our people identified one of those "high-impact" hazards I talked about in the last chapter, his or her partner tried to find something bigger. And it snowballed.

I've already mentioned that the site saw a radical change in the number of incidents (over 50%) within the span of one year after committing to the ICE method. It would be easy to write off the work put in addressing the hazards and say that number was pure coincidence. I'm sure there are many who would. There is no denying, however, that for every potential hazard removed, the potential for injury it carried with it was gone as well. The safety paradox will never go away, but you can't argue fact.

Shortly before I left that project, I was approached by a carpenter. On a site with nearly 2,000 workers, I'd never met nor seen him before, but he had known who I was. He stopped me and put out his hand. I shook it and he explained why he had sought me out.

"I've been working in this area for more than 25 years," he said. "This is the safest job I've ever been on. Thank you for that."

I was dumbfounded but I graciously accepted his gratitude. "Just trying to do what I can," I said. "Thank you for telling me that, it means a lot."

Any safety practitioner who has lived a success like that knows that's where true professional satisfaction comes from. You can create the best charts and graphs the world has ever seen and submit reports to corporate that make you look like a safety god. But none of that means anything if you don't have an impact on the people.

So that's it. Build a foundation, train like you mean it, identify your greatest risks, then attack the hazards. I'm not sure why anyone has ever thought safety management *needs* to be more complicated than that. Those four things are more than enough to tackle. But don't take my word for it. Get to work and find out for yourself.

EPILOGUE
One Night at a Hometown Diner

Nick and I met for dinner one night in November in 2015. We talked for hours about all that had happened since the last time we had seen each other, but we didn't reminisce. We didn't need to.

I noticed he was much frailer than before and his giant frame had begun to crumple in on itself. I listened as he told me the last story he would ever tell me. It was one I won't recount here in its entirety because it's his story, not mine. But it was as engaging as any he had ever told. It was one I had wished to be a part of, but knowing now how it ended I know why he didn't want me to join in.

His last job had been a consultation job in Libya in 2012. I remember having begged him to take me with him (something he certainly had the clout to do) so I could gain the kind of international experience he had always spoken of in such high regard. He had been adamant that it "wasn't the job" for me at the time. Anyone who is familiar with what occurred in that country in September of that year will probably be able to put the pieces together. It was not a good place for Americans at that time. He had been completely right to keep me out of it.

We finished our dinner, complete with chili and cornbread, and talked until the staff awkwardly stood in the corner waiting for us to leave. He and I walked to his tricked out green pickup that was far too hip for his age and said our goodbyes, looking forward to the next time I came to town.

My phone rang several months later. Nick's name and picture lit up the screen and I answered excitedly.

"Hey Nick, what's up?" For a moment I only heard silence.

"Jason," I knew instantly. "Nick passed away yesterday." His daughter's words still echo in my head. I don't know how that conversation ended.

I hope that this book keeps his stories, and their lessons, alive. Help me use them to keep up the pursuit of #relentlesssafety.

WORKS CITED

Blake McGowan, "Workplace Safety Programs: Are They Worth $400,000 Per Year?" *Professional Safety*, December 2018, pp. 46–47.

Occupational Health and Safety Administration, 2002. OSHA's No-Fault Recordkeeping System Requires Recording Work-Related Injuries and Illnesses, Regardless of the Level of Employer Control or Non-Control Involved. www.osha.gov/laws-regs/standardinterpretations/2002-02-06.

Margaret Venuto, Lorie C. Brosch, Juste Tchjanda and Thomas Leo Cropper, "Retrospective Case Series of Five Nontraumatic Deaths among U.S. Air Force Basic Military Trainees (1997–2007)" *Military Medicine*, Vol. 176, August 2011, p. 940.

Index

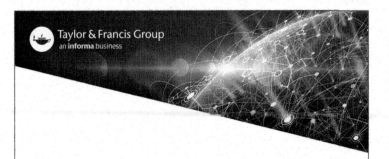